Advances in Combinatorial Mathematics

Ilias S. Kotsireas • Eugene V. Zima
Editors

Advances in Combinatorial Mathematics

Proceedings of the Waterloo Workshop in Computer Algebra 2008

Springer

Editors
Ilias S. Kotsireas
Eugene V. Zima
Wilfrid Laurier University
Department of Physics and Computer Science
75 University Avenue West
Waterloo, Ontario N2L 3C5
Canada
ikotsire@wlu.ca
ezima@wlu.ca

ISBN 978-3-642-03561-6 e-ISBN 978-3-642-03562-3
DOI 10.1007/978-3-642-03562-3
Springer Heidelberg Dordrecht London New York

Library of Congress Control Number: 2009938633

Mathematics Subject Classification (2000): 05-xx, 15-xx

Cover design: WMXDesign GmbH, Heidelberg

Printed on acid-free paper

Springer is part of Springer Science+Business Media (www.springer.com)

This book is dedicated to the 70th birthday of Georgy Egorychev, the author of the influential, milestone book "Integral Representation and the Computation of Combinatorial Sums", and a recipient of the Fulkerson Prize for solving the van der Waerden conjecture on the determination of the minimum of the permanent of a doubly stochastic matrix.

Foreword

It is a pleasure for me to have the opportunity to write the foreword to this volume, which is dedicated to Professor Georgy Egorychev on the occasion of his seventieth birthday. I have learned a great deal from his creative and important work, as has the whole world of mathematics. From his life's work (so far) in having made distinguished contributions to fields as diverse as the theory of permanents, Lie groups, combinatorial identities, the Jacobian conjecture, etc., let me comment on just two of the most important of his research areas.

The *permanent* of an $n \times n$ matrix A is

$$\mathrm{Per}(A) = \sum a_{1,i_1} a_{2,i_2} \dots a_{n,i_n}, \tag{1}$$

extended over the $n!$ permutations $\{i_1, \dots, i_n\}$ of $\{1, 2, \dots, n\}$. Thus, the permanent is "like the determinant except for dropping the sign factors from the terms." However by dropping those signs, one loses almost all of the friendly characteristics of determinants, such as the fact that $\det(AB) = \det(A)\det(B)$, the invariance under elementary row and column operations, and so forth. The permanent is a creature of multilinear algebra, rather than of linear algebra, and is much crankier to deal with in virtually all of its aspects, both theoretical and algorithmic.

Nonetheless the permanent is quite an important concept, for example in combinatorial mathematics. The permanent of a matrix whose entries are all either 0's or 1's is (see (1) above) the number of permutations of n letters for which all n of the entries $\{a_{\nu,i_\nu}\}_{\nu=1}^{n}$ are 1's, and this is a valuable tool for counting permutations with restricted positions, for counting Latin rectangles and squares, and so forth.

In 1926, B. L. van der Waerden conjectured that among all $n \times n$ matrices whose entries are nonnegative real numbers and whose row and column sums are all equal to 1, the matrix whose permanent is as small as possible is uniquely the one whose entries are all equal to $1/n$. In view of the numerous applications of permanents, the truth of this conjecture would have valuable consequences. Fifty-six years later the conjecture was proved [1] by Egorychev. (Another proof, found almost simultaneously, is due to Falikman [4].)

That achievement alone would have been enough to assure Professor Egorychev's place on the honor roll of great mathematicians, but we must mention another aspect of his research that reinforces this evaluation. I am referring to his work on combinatorial identities, as described in his book ([2], [3]). There he has shown how a wide class of combinatorial identities can be proved and/or discovered by the methods of complex analysis, thereby making an important contribution to the unity of a subject which has in the past been highly fragmented, but which now, thanks to his and other remarkable advances, is starting to show signs of maturity. Professor Egorychev discusses some recent developments of this line of thought in Chapter 1 below.

As you read the contributions by his friends and colleagues in this volume, take note of the variety and the beauty of the fields of mathematics that they encompass, and reflect on the varied and extensive advances in mathematics that we owe to Professor Georgy Egorychev.

References

[1] Егорычев, Г. П. [Egorychev, G. P.] Решение проблемы ван-дер-Вардена для перманентов. (Russian) [Solution of the van der Waerden problem for permanents] Preprint IFSO-13M. *Akad. Nauk SSSR Sibirsk. Otdel., Inst. Fiz.*, Krasnoyarsk, 1980. 12 pp.

[2] Егорычев, Г. П. [Egorychev, G. P.], Интегральное представление и вычисление комбинаторных сумм. (Russian) [Integral representation and computation of combinatorial sums] Izdat. "Nauka" Sibirsk. Otdel., Novosibirsk, 1977. 283 pp.

[3] Egorychev, G. P. Integral representation and the computation of combinatorial sums. Translated from the Russian by H. H. McFadden. Translation edited by Lev J. Leifman. Translations of Mathematical Monographs, **59**. American Mathematical Society, Providence, RI, 1984. x+286 pp.

[4] Фаликман Д. И. [Falikman, D. I.], Доказательство гипотезы ван дер Вардена о перманенте дважды стохастическои матрицы [Proof of the van der Waerden conjecture on the permanent of a doubly stochastic matrix.] (Russian) *Mat. Zametki* 29 (1981), no. 6, 931–938, 957.

Philadelphia, PA, USA April 23, 2009 *Herbert S. Wilf*

Preface

The Second Waterloo Workshop on Computer Algebra (WWCA 2008) was held May 5-7, 2008 at Wilfrid Laurier University, Waterloo, Canada. This conference was dedicated to the 70th birthday of Georgy Egorychev (Krasnoyarsk, Russia), who is well known and highly regarded as the author of the influential, milestone book “Integral Representation and the Computation of Combinatorial Sums,” which described a regular approach to combinatorial summation, today also known as the method of coefficients. Another great success of this Russian mathematician came in 1980, when he solved the van der Waerden conjecture on the determination of the minimum of the permanent of a doubly stochastic matrix and was awarded the D.R. Fulkerson Prize.

Topics discussed at the workshop[1] were devoted to these two themes (combinatorial and algorithmic summation and special polynomials) and related problems in enumerative combinatorics. The workshop’s format included invited lectures and presentations, and it attracted international participants from the USA, Europe, Taiwan, as well as several Canadian universities. Different aspects of the method of coefficients and its relation to algorithmic summation methods and methods of proving combinatorial identities were thoroughly discussed by George E. Andrews (Pennsylvania State University, USA), Georgy Egorychev (Siberian Federal University, Russia), Ira Gessel (Brandeis University, USA), I-Chiau Huang (Institute of Mathematics, Taiwan), Peter Paule (RISC-Linz, Austria), Marko Petkovsek (University of Ljubljana, Slovenia), and Doron Zeilberger (Rutgers University, USA). The theory and applications of the permanent and other special polynomials were presented by Leonid Gurvits (Los Alamos National Laboratory, USA) and Herbert Wilf (University of Pennsylvania, USA). Michiel Hazewinkel (CWI, the Netherlands) discussed the “niceness” of mathematical objects and theorems.

The workshop was financially supported by the Fields Institute of the University of Toronto and various offices of Wilfrid Laurier University.

---

[1] http://www.cargo.wlu.ca/wwca2008/

This book presents a collection of selected formally refereed papers submitted after the workshop. The topics discussed in this book are closely related to Georgy Egorychev's influential works.

This book would not have been possible without the dedication and hard work of the anonymous referees, who supplied detailed referee reports and helped authors to improve their papers significantly. Finally, we wish to thank the people at Springer-Verlag, in particular Ruth Allewelt and Martin Peters, for working closely with us and for their unequivocal support throughout the entire publication process.

Waterloo,
May 2009

Ilias S. Kotsireas
Eugene V. Zima

Contents

Chapter 1
Method of Coefficients: an algebraic characterization and recent applications

Georgy P. Egorychev

Abstract The article is devoted to the algebraic-logical foundations of the analytic approach to summation problems in various fields of mathematics and its applications. Here we present the foundations of the method of coefficients developed by the author in late 1970's and its recent applications to several well-known problems.

1.1 Introduction

The article is devoted to the algebraic-logical foundations of the analytical approach to summation problems in various fields of mathematics and its applications. Here we present the foundations of the method of integral representations and computation of combinatorial sums (the method of coefficients) developed by the author in the end of 1970's [25] and its recent applications to several well-known problems (see reviews in [26, 31]). The article contains several new results, including the method of coefficients (the set of inference rules and the Completeness Lemma) with operations in the ring of formal Dirichlet series of usual type, as well as several new properties of the characteristic function of the stopping height for the Collatz problem [27, 28], and the solutions of two interesting problems of summation in the theory of holomorphic functions in $\mathbb{C}^n$. Finally we shall give a new algebraic characterization of the method of coefficients, which is based on the *φ-operation of isomorphism* [9, 20, 59], generated by the classical one-to-one mapping φ between the set $\mathscr{A}$ of numerical sequences and the set $\mathscr{B}$ of generating series of given type. These results allow one to formulate the following statement [32].

E-principle of summation: *each pair of inverse linear transforms (for sequences, series, functions, etc.), independently of the way of definition of the one-to-one mapping φ, generates the corresponding method of summation (the method of coefficients).*

Georgy P. Egorychev
Siberian Federal University, Krasnoyarsk, RUSSIA, e-mail: anott@scn.ru

I.S. Kotsireas, E.V. Zima (eds.), *Advances in Combinatorial Mathematics*,
DOI 10.1007/978-3-642-03562-3_1,

This principle provides for the first time a foundation for the classical method of generating functions (generating integrals) as a method of summation for different classes of generating series (the Completeness Lemma). It also makes it possible to reduce the variety of calculations with them to a uniform combinatorial scheme, and to set a new extensive program of solving open summation problems.

1.2 The method of generating functions as a method of summation (the method of coefficients)

1.2.1 Computational scheme

The general scheme of the method of integral representations of sums can be broken down into the following steps [25].

1. *Assignment of a table of integral representations of combinatorial numbers. For example, the binomial coefficients* $\binom{n}{k}$, n, $k = 0, 1, \ldots$,

$$\binom{n}{k} = \mathbf{res}_w\,(1+w)^n\,w^{-k-1} = \frac{1}{2\pi i}\int_{|w|=\rho}(1+w)^n\,w^{-k-1}dw,\ \rho > 0; \qquad (1.1)$$

$$\binom{n+k-1}{k} = \mathbf{res}_w\,(1-w)^{-n}\,w^{-k-1} = \frac{1}{2\pi i}\int_{|w|=\rho}(1-w)^{-n}\,w^{-k-1}dw,\ 0 < \rho < 1. \qquad (1.2)$$

Stirling numbers of the second kind $S_2(n,k)$, n, $k = 0, 1, \ldots$([25], p. 273): $S_2(0,0) := 1$, *and*

$$S_2(n,k) = \mathbf{res}_w\{(-1+\exp w)^n w^{-k-1}\} = \frac{1}{2\pi i}\int_{|w|=\rho}(-1+\exp w)^n w^{-k-1}dw,\ \rho > 0. \qquad (1.3)$$

The Kronecker symbol $\delta(n,k)$, n, $k = 0, 1, \ldots$,

$$\delta(n,k) = \mathbf{res}_w w^{-n+k-1}. \qquad (1.4)$$

2. *Representation of the summand* a_k *of the original sum* $\sum_k a_k$ *by a sum of product of combinatorial numbers.*

3. *Replacement of the combinatorial numbers by their integrals.*

4. *Reduction of products of integrals to multiple integral.*

5. *Interchange of the order of summation and integration. This gives the integral representation of original sum with the kernel represented by a series. The use of this transformation requires us to deform the domain of integration in such a way as to obtain the series under the integral which converges uniformly on this domain saving the value of the integral.*

7. *Summation of the series under the integral sign. As a rule, this series turns out to be a geometric progression* [46]. *This gives the integral representation of the original sum with the kernel in closed form.*

7. *Computation of the resulting integral by means of tables of integrals, iterated integration, the theory of one-dimensional and multidimensional residues, or other suitable methods.*

1.2.2 Operations with formal power series and the inference rules

Hans Rademacher [87] has noted, that the applications of the method of generating functions is connected usually with use of operations over the Laurent series and the Dirichlet series. Earlier the author has developed the method of integral representations and calculation of combinatorial sums of various types [25, 26, 29, 31], connected with use of the theory of analytic functions, the theory of multiple residues in $\mathbb{C}^n$ and the formal power Laurent series over $\mathbb{C}$. In this section we give an analogous construction and the foundation of the method of coefficients for classic formal Dirichlet series of one variable over $\mathbb{C}$.

1.2.2.1 Laurent power series: definition and properties of the residue operator

Using the **res** concept and its properties the idea of integral representations can be extended on sums that allow computation with the help of formal Laurent power series of one and several variables over $\mathbb{C}$. The **res** concept is directly connected with the classic concept of residue in the theory of analytic functions and which may be used with series of various types. This connection has enabled us to express properties of **res** operator analogous to properties of residue in the theory of analytic functions. This in turn allows us to unify the scheme of the method of integral representations independently of what kind of series – convergent or formal – is being used (separately, or jointly) in the process of computation of a particular sum.

In this section we shall restrict ourselves to explaining only one-dimensional case, although in further computations the **res** concept shall also be used for multivariate series. Besides, the one-dimensional case is interesting by itself in the computation of multiple integrals in terms of repeated integrals.

Let L be the set of formal Laurent power series over $\mathbb{C}$ containing only finitely many terms with negative degrees. The *order* of the monomial $c_k w^k$ is k. The *order* of the series $C(w) = \sum_k c_k w^k$ from L is the minimal order of monomials with nonzero coefficient. Let L_k denote the set of series of order k, $L = U_{k=-\infty}^{\infty} L_k$. Two series $A(w) = \sum_k a_k w^k$ and $B(w) = \sum_k b_k w^k$ from L are equal iff $a_k = b_k$ for all k. We can introduce in L operations of addition, multiplication, substitution, inversion and differentiation [15, 35, 47]. The ring L is a field [85]. Let $f(w)$, $\psi(w) \in L_0$. Below we shall use the following notations: $h(w) = wf(w) \in L_1$, $l(w) = w/\psi(w) \in L_1$, $z'(w) = \frac{d}{dw}z(w)$, $\overline{h} = \overline{h}(z) \in L_1$ – the inverse series of the series $z = h(w) \in L_1$.

For $C(w) \in L$ define the *formal residue* as

$$\mathbf{res}_w C(w) = c_{-1}. \tag{1.5}$$

Let $A(w) = \sum_k a_k w^k$ be the *generating function* for the sequence $\{a_k\}$. Then

$$a_k = \mathbf{res}_w A(w) w^{-k-1}, \ k = 0, 1, \ldots. \tag{1.6}$$

For example, one of the possible representations of the binomial coefficient is

$$\binom{n}{k} = \mathbf{res}_w \, (1+w)^n \, w^{-k-1}, \ k = 0, 1, \ldots, n. \tag{1.7}$$

There are several properties (*inference rules*) for the **res** operator which immediately follow from its definition and properties of operations in formal Laurent power series over $\mathbb{C}$. We list only a few of them which will be used in this article. Let $A(w) = \sum_k a_k w^k$ and $B(w) = \sum_k b_k w^k$ be generating functions from L.
Rule 1 (Removal).

$$\mathbf{res}_w A(w) w^{-k-1} = \mathbf{res}_w B(w) w^{-k-1} \text{ for all } k \text{ iff } A(w) = B(w). \tag{1.8}$$

Rule 2 (Linearity). *For any α, β from $\mathbb{C}$*

$$\alpha \ \mathbf{res}_w A(w) w^{-k-1} + \beta \ \mathbf{res}_w B(w) w^{-k-1} = \mathbf{res}_w((\alpha A(w) + \beta B(w)) w^{-k-1}). \tag{1.9}$$

By induction from (1.9) *it follows, that the operators $\sum$ and* **res** *are commutative.*
Rule 3 (Substitution). a) *For $w \in L_k$ $(k \geq 1)$ and $A(w)$ any element of L, or* b) *for $A(w)$ polynomial and w any element of L including a constant*

$$\sum_k w^k \mathbf{res}_z \left(A(z) z^{-k-1} \right) = [A(z)]_{z=w} = A(w). \tag{1.10}$$

Rule 4 (Inversion). *For $f(w)$ from L_0*

$$\sum_k z^k \mathbf{res}_w \left(A(w) f(w)^k w^{-k-1} \right) = \left[A(w)/f(w) h'(w) \right]_{w=\overline{h(z)}}, \tag{1.11}$$

where $z = h(w) = w f(w) \in L_1$.
Rule 5 (Change of variables). *If $f(w) \in L_0$, then*

$$\mathbf{res}_w \left(A(w) f(w)^k w^{-k-1} \right) = \mathbf{res}_z (\left[A(w)/f(w) h'(w) \right]_{w=\overline{h(z)}} z^{-k-1}), \tag{1.12}$$

where $z = h(w) = w f(w) \in L_1$.
Rule 6 (Differentiation).

$$k \ \mathbf{res}_w A(w) w^{-k-1} = \mathbf{res}_w A'^{-k}. \tag{1.13}$$

1.2.2.2 Dirichlet series: definitions and properties of the $[q^{-s}]$ operator

Let H be the set of formal Dirichlet series $A(s) = \sum_{k\geq 1} a_k k^{-s}$ of usual type in formal variable s with complex coefficients. Two series $A(s) = \sum_k a_k k^{-s}$ and $B(s) = \sum_k b_k k^{-s}$ from H are equal iff $a_k = b_k$ for all k. We can introduce in H operations of addition, multiplication and differentiation of series [63, 70]. The set H is a ring.

Let G be the set of formal exponential series of type $A(s) = \sum_{q\in\mathbb{Q}} a_q q^{-s}$ in variable s with complex coefficients, $H \subset G$. For $A(s) \in G$ define *the* $[q^{-s}]$*-operator* as

$$a_q = \left[q^{-s}\right](A(s)),\ \forall q \in \mathbb{Q}, \tag{1.14}$$

i.e. the $[q^{-s}]$-operator is the coefficient at the exponent q^{-s} of the series $A(s)$. If $A(s) = \sum_k a_k k^{-s}$ from H is the generating function for the sequence $\{a_k\}$ then as usual

$$a_k = \left[k^{-s}\right](A(s)),\ k = 1,2,\ldots \tag{1.15}$$

Remark. Here the sign $\sum_{q\in\mathbb{Q}}$ is analogous to the sign $\sum_{k\in\mathbb{N}}$ which we often use instead the sign $\sum_{k=0}^{\infty}$ for power series and formal Dirichlet series of usual type (see also [85], p.118). The notion of the formal exponential series $A(s) = \sum_{q\in\mathbb{Q}} a_q q^{-s}$ from G is necessary below in the proof of formulae in section 1.2.3.

For example, we have the following representation for the coefficients of zeta-function $\zeta(s) := \sum_{k\geq 1} k^{-s}$, and the inverse of it $1/\zeta(s) = \sum_{k\geq 1} \mu(k) k^{-s}$, $\operatorname{Re} s > -1$:

$$1 = \left[k^{-s}\right](\zeta(s)),\ k = 1,2,\ldots, \tag{1.16}$$

$$\mu(k) = \left[k^{-s}\right](1/\zeta(s)),\ k = 1,2,\ldots, \tag{1.17}$$

where μ is the Möbius function.

There are several properties (*inference rules*) for the $[q^{-s}]$-operator which immediately follow from its definition and properties of operations on the formal Dirichlet series over $\mathbb{C}$. Let $A(s) = \sum_k a_k k^{-s}$ and $B(s) = \sum_k b_k k^{-s}$ be the generating functions for the sequences $\{a_k\}$ and $\{b_k\}$from H.
Rule 1 (Removal).

$$\left[k^{-s}\right](A(s)) = \left[k^{-s}\right](B(s)) \text{ for all } k \text{ iff } A(s) = B(s). \tag{1.18}$$

Rule 2 (Shifting). For any $d, n \in \mathbb{N}$

$$[(n/d)^{-s}](A(s)) = [n^{-s}](d^{-s}A(s)). \tag{1.19}$$

Rule 3 (Linearity). *For any* α, β *from* $\mathbb{C}$

$$\alpha \left[q^{-s}\right](A(s)) + \beta \left[q^{-s}\right](B(s)) = \left[q^{-s}\right](\alpha A(s) + \beta B(s)). \tag{1.20}$$

By induction from (1.20) *follows, that operators* $\sum$ *and* $[q^{-s}]$ *commute.*
Rule 4 (Substitution).

$$\sum_{k\geq 1} k^{-s} \left[k^{-t}\right](A(t)) = (A(t))|_{t=s} = A(s). \tag{1.21}$$

Rule 5 (Differentiation).

$$\left[k^{-s}\right](A'(s)) = -\ln k \times \left[k^{-s}\right](A(s)), k = 1,2,.... \tag{1.22}$$

1.2.3 The problem of completeness

1.2.3.1 Statement of the problem

In solving analytic problems with the help of generating functions we usually encounter one of the following interconnected problems.

Problem A. *Suppose that a series $S(w) = \sum_k s_k w^k$ from L is expressed in terms of the series $A(w) = \sum_k a_k w^k$, $B(w) = \sum_k b_k w^k$,..., $D(w) = \sum_k d_k w^k$ from L with the help of different operations on the formal Laurent power series over $\mathbb{C}$, i.e. the formula*

$$S(w) = F(A(w), B(w), \dots, D(w)) \tag{1.23}$$

is given. For each k find the formula

$$s_k = f(\{a_k\}, \{b_k\}, \dots, \{d_k\}) \tag{1.24}$$

for the terms of sequence $\{s_k\}$ as a function of the terms of sequences $\{a_k\}$, $\{b_k\}, \dots, \{d_k\}$.

Definition. *A sequence $\{s_k\}$ is called of A-type with respect to terms of sequences $\{a_k\}, \{b_k\}, \dots, \{d_k\}$, if it is determined by a formula of type* (1.24).

Problem B. *Let for each k the formula $s_k = f(\{a_k\}, \{b_k\}, \dots, \{d_k\})$, $\forall k = 0, 1, \cdots$, with respect to terms of number sequences $\{a_k\}, \{b_k\}, \dots, \{d_k\}$ be given, but a functional dependence* (1.23) *between its generating functions is unknown. It is required to find out, whether the initial formula $s_k = f(\{a_k\}, \{b_k\}, \dots, \{d_k\})$ is a formula of A-type, and if yes, then to find formula $S(w) = F(A(w), B(w), \dots, D(w))$.*

Definition. *A set of rules for* **res** *operator ($[q^{-s}]$-operator) is called complete, if it allows one to solve problem* ***B***.

1.2.3.2 Completeness Lemma: Laurent and Dirichlet series

Completeness Lemma.

(a) *The set of rules 1 – 6 for the* **res** *operator of the formal Laurent series is complete* [26].

(b) *The set of rules 1 – 5 for the $[q^{-s}]$-operator of the formal Dirichlet series of usual type is complete.*

Proof.

(a) In [25] (pp. 31–35) and [26] we use induction on the number of different operations over sequences $\{a_k\}$, $\{b_k\}$, ..., $\{d_k\}$ in (1.24) generating the given sequence $\{s_k\}$. On the first step of induction a series $S(w)$ is obtained with the help of series $A(w)$ and $B(w)$ from L by one operation over formal Laurent power series (addition, multiplication, etc.).

(b) Below we perform analogous calculations for the formal Dirichlet series of usual type. On the first step of induction a series $S(s)$ is obtained with the help of the formal Dirichlet series $A(s)$ and $B(s)$ from H and one of the operations of addition and multiplication. We should give the solution to recursive relations that corresponds to each of these operations.

Addition operation. If $c_k = a_k + b_k$, $k = 1,2,...$, then by formulae (1.15) for the coefficients c_k, a_k and b_k we obtain

$$\left[k^{-s}\right](C(s)) = \left[k^{-s}\right](A(s)) + \left[k^{-s}\right](B(s)),\ k = 1,2,...,$$

(by the linearity rule and the removal rule)

$$\Leftrightarrow \left[k^{-s}\right](C(s)) = \left[k^{-s}\right](A(s) + (B(s)) \text{ for all } k \Leftrightarrow C(s) = A(s) + B(s).$$

Multiplication operation. On one hand we have $C(s) = A(s) \times B(s) := \sum_k c_k k^{-s}$, where

$$c_k = \sum_{d|k} a_d b_{k/d},\ k = 1,2,..., \tag{1.25}$$

where (and up to the end of the section) the summation is over all the divisors d of natural number k. Conversely, if the identity (1.25) holds, then for $k = 1,2,...,$ we get:

$$c_k = \sum_{d|k} a_d b_{k/d},$$

(the change of coefficients a_d and $b_{k/d}$ by formulae (1.15))

$$\sum_{d|k} \left[d^{-t}\right](A(t)) \times \left[(k/d)^{-s}\right](B(s)) = \sum_{d=1}^{\infty} \left[d^{-t}\right](A(t)) \times \left[(k/d)^{-s}\right](B(s))$$

(as added terms are equal to zero by the definition (1.14) of the $\left[q^{-s}\right]$-operator, and further the shifting rule over s)

$$= \sum_{d=1}^{\infty} \left[d^{-t}\right]\left\{\left[k^{-s}\right](d^{-s}A(t)B(s))\right\}\ldots$$

(interchanging the order of $\sum$ and $\left[d^{-t}\right]\left[k^{-s}\right]$ and splitting the sum over the index d)

$$= \left[k^{-s}\right]\left(B(s) \times \left\{\sum_{d=1}^{\infty} d^{-s}\left[d^{-t}\right](A(t))\right\}\right)$$

(the substitution rule for an expression in braces and the change $t = s$)

$$= \left[k^{-s}\right]\{B(s) \times (A(t))|_{t=s}\} = \left[k^{-s}\right]\{B(s)A(s)\}.$$

Now by (1.25) we have

$$c_k := \left[k^{-s}\right](C(s)) = \left[k^{-s}\right]\{B(s)A(s)\},\ k = 1,2,...,$$

and the removal rule of the $[k^{-s}]$-operator gives us the required formula

$$C(s) = B(s)A(s).$$

If the hypothesis of Lemma holds for $n-1$ operations, then the next inductive step is similar to the initial step.

In the following illustrative example we use only concepts and the inference rules for the formal Dirichlet series.

Example. The celebrated *Möbius inversion formula* states that

$$f(n) = \sum_{d|n} g(d),\ n = 1,2,\ldots \Leftrightarrow g(n) = \sum_{d|n} \mu(d) f(n/d),\ n = 1,2,\ldots. \quad (1.26)$$

Proof. Let $F(s) = \sum_{n\geq 1} f(n) n^{-s}$ and $G(s) = \sum_{n\geq 1} g(n) n^{-s}$ from H be the generating functions for the sequences $\{f(n)\}$ and $\{g(n)\}$. Repeating the same scheme of calculations we get:

$$g(n) := [n^{-s}](G(s)) = \sum_{d|n} \mu(d) f(n/d)$$

(the substitution using (1.17) and (1.15): $f(n/d) = [(n/d)^{-s}](F(s))$ and $\mu(d) = [d^{-t}](1/\zeta(t))$)

$$= \sum_{d|n} [d^{-t}](1/\zeta(t)) \times [(n/d)^{-s}](F(s)) = \sum_{d|n} \left[d^{-t}\right](1/\zeta(t)) \times \left[n^{-s}\right](d^{-s}F(s))$$

$$= \sum_{d\geq 1} \ldots = \left[n^{-s}\right]\left(F(s) \times \left\{\sum_{d=1}^{\infty} d^{-s}\left[d^{-t}\right](1/\zeta(t))\right\}\right)$$

$$= \left[n^{-s}\right](F(s) \times (1/\zeta(t))|_{t=s}) = \left[n^{-s}\right](F(s)/\zeta(s)).$$

Thus we obtain

$$[n^{-s}](G(s)) = \left[n^{-s}\right](F(s)/\zeta(s)),\ \text{for all } n,$$

and the removal rule of the $[n^{-s}]$-operator gives us

$$G(s) = F(s)/\zeta(s) \Leftrightarrow F(s) = \zeta(s) G(s) = \sum_{k\geq 1} k^{-s} \times \sum_{k\geq 1} g_k k^{-s},$$

i.e.

$$f(n) = \sum_{d|n} g(d),\ n = 1,2,\ldots.$$

Remark. *Completeness Lemma supports the possibility of finding with the help of the method of coefficients an operational (integral) representation for those sums, which admit the calculation with formal Laurent power series and Dirichlet formal series with complex coefficients. Basic difficulty in the use of this method (the set of inference rules and the Completeness Lemma) consists in the solution of problems of classification and recognition of expressions of A-type, and in construction of algorithms of induction search though these problems have found the successful solution in many concrete cases of calculation of combinatorial sums* [25].

1.2.4 Connection with the theory of analytic functions

If a formal power series $A(w) \in L$ converges in a punctured neighborhood of zero, then the definition of $\mathbf{res}_w A(w)$ coincides with the usual definition of $\mathbf{res}_{w=0} A(w)$, used in the theory of analytic functions. The formula (1.6) is an analog of the well-known integral Cauchy formula

$$a_k = \frac{1}{2\pi i} \oint_{|w|=\rho} A(w) w^{-k-1} dw$$

for the coefficients of the Taylor series in a punctured neighborhood of zero. The substitution rule (1.10) of the **res** operator is a direct analog of the famous Cauchy theorem. Similarly, it is possible to introduce the definition of formal residue at the point of infinity, the logarithmic residue and the theorem of residues (all necessary concepts and results in the theory of residues in one and several complex variables, see [2, 25, 34, 76, 95, 107]). Moreover, it is easy to see that each rule of the **res** operator can be simply proven by reduction to the known formula in the theory of residues for corresponding rational function [25].

The theory of Dirichlet series of usual type can be found in many books on the theory of holomorphic functions and analytical number theory (see, for example, [63, 70]).

1.3 Several recent applications

1.3.1 The characteristic function of the stopping height for the Collatz conjecture

The $3x+1$ problem is known under different names. It is often called *Collatz problem, Ulam problem, the Syracuse problem, Kakutani problem, and Hasse algorithm* [60]. Consider the sequence of iterations $(n, f(n), f(f(n)), \ldots)$, where

$$f(n)=\begin{cases}(3n+1)/2, & \text{for odd } n,\\ n/2, & \text{for even } n.\end{cases} \tag{1.27}$$

The $3x+1$ conjecture states that for any natural number n this sequence will contain the number 1. The index of the first element equal to 1 in this sequence is called *stopping height* of the instance of Collatz problem and is denoted $\sigma(n)$.

The following arithmetic reformulation of the Collatz problem is given in [71].

Theorem 1.[1] *The $3x+1$ conjecture is true iff for every positive integer a there are natural numbers w and v such that $a\le w$ and*

$$\binom{2w+1}{w}\binom{4(w+1)v+1}{v}\sum_{r=0}^{\infty}\sum_{s=0}^{\infty}\sum_{t=0}^{\infty}\binom{v}{r}\binom{w(v-r)}{s}\binom{wr}{t}\times \tag{1.28}$$
$$\binom{2s+2t+r+(4w+3)v+1}{3((4w+4)t+a)+2(4w+4)r+(4w+4)s}\times$$
$$\binom{3((4w+4)t+a)+2(4w+4)r+(4w+4)s}{2s+2t+r+(4w+3)v+1}\equiv 1 \pmod 2.$$

In [27, 31] one can find the following reformulation of (1.28) obtained with the help of the method of coefficients and based on congruences (modulo 2)

$$(1+u)^{\alpha}\equiv 1+u^{\alpha},\ \ (1+u)^{\alpha-1}\equiv\sum_{s=0}^{\alpha-1}u^{s},\ \ (1-(\alpha-1)^2u)^{-1/(\alpha-1)}\equiv\prod_{s=0}^{\infty}\left(1+u^{\alpha^{s}}\right), \tag{1.29}$$

where $\alpha=2^{x}$, $x\in\mathbb{N}$:

Let $a,v,w\in\mathbb{N}$ and denote

$$S=\sum_{r=0}^{\infty}\sum_{s=0}^{\infty}\sum_{t=0}^{\infty}\binom{v}{r}\binom{w(v-r)}{s}\binom{wr}{t}\binom{2s+2t+r+(4w+3)v+1}{3(4w+4)t+2(4w+4)r+(4w+4)s+a}\times$$
$$\binom{3(4w+4)t+2(4w+4)r+(4w+4)s+a}{2s+2t+r+(4w+3)v+1}. \tag{1.30}$$

Then

$$S=\mathbf{res}_u\{g(u)\,u^{-(4w+3)v+a-2}\}, \tag{1.31}$$

where

$$g(u)=\left(\left(1+u^{-2+(4w+4)}\right)^{w}+u^{-1+2(4w+4)}\left(1+u^{-2+3(4w+4)}\right)^{w}\right)^{v}. \tag{1.32}$$

This leads to the following reformulation of $3x+1$ conjecture.

Theorem 2 [27]. *The $3x+1$ conjecture is true iff for every positive integer a there are natural numbers r and $\alpha=2^{x+2}$, where $x\in\mathbb{N}$, such that $a\le -1+\alpha/4$,*

---

[1] Careful investigation of this result along with computer experiments shows that this formula and analogous statements ([71], Theorem 1, Corollary 1 – 3) are not valid. The following correction is required: the term a has to be replaced by $a/3$ in order to make it work. We shall use the corrected version of (1.28) below.

and the following congruence is true

$$\mathbf{res}_u u^{-\alpha^r+a-1}\prod_{t=0}^{\infty}\left(\sum_{s=0}^{-1+\alpha/4}\left(u^{s(-2+\alpha)\alpha^t}+u^{(-1+2\alpha+s(-2+3\alpha))\alpha^t}\right)\right)\equiv 1 \pmod 2, \tag{1.33}$$

1.3.1.1 Properties of the characteristic function of the stopping height

Definition. *In accordance with* (1.33) *denote*

$$Q_\alpha(u) := d_\alpha(u)\prod_{t=1}^{\infty} d_\alpha(u^{\alpha^t}), \tag{1.34}$$

where the polynomial

$$d_\alpha(u) = 1+u^{-1+2\alpha}+\sum_{s=1}^{-1+\alpha/4}\left(u^{s(-2+\alpha)}+u^{-1+2\alpha+s(-2+3\alpha)}\right).$$

It is shown in [27], that the coefficients of this formal power series $Q_\alpha(u)$ over integers

$$Q_\alpha(u) = \sum_k q_k(\alpha)u^k \tag{1.35}$$

are equal to either 0 or 1. Therefore, the congruence (1.33) is a theoretical-functional reformulation of the Collatz conjecture. It was noted in [27, 28], that the parameter r in Theorem 2 is equal to the stopping height $\sigma(n)$. Thus under the assumptions of Theorem 2, now the equivalent formulation of the Collatz conjecture can be given by the equality

$$q_{-n+\alpha^{\sigma(n)}} = 1. \tag{1.36}$$

The last formulation is more attractive than (1.28), and these properties of the function $Q_\alpha(u)$ allows us to call it *the characteristic function of the stopping height in the Collatz conjecture.*

Lemma (*Characteristic property*) *For any α the coefficients of the formal power series $Q_\alpha(u)\in H(\mathbb{Z})$ in* (1.35) *are equal to either* 0 or 1.

Proof. The statement of Lemma was proven in [27] only for $k=\alpha^q-n,\ n\in\mathbb{N}$. However, that proof can be repeated for an arbitrary k.

Lemma (*Functional equations*) *For any α, the function $Q_\alpha(u)$ is uniquely defined by the functional equation*

$$Q_\alpha(0)=1,\quad Q_\alpha(u)=d_\alpha(u)Q_\alpha(u^\alpha). \tag{1.37}$$

The following congruence holds

$$(d_\alpha(u))^{-1/(\alpha-1)} \equiv Q_\alpha(u) (\text{mod } 2), \tag{1.38}$$

where in accordance with (1.34) *series* $g_\alpha(u) = (d_\alpha(u))^{-1/(\alpha-1)} \in H(\mathbb{Q})$, *and the equation* $d_\alpha(u) \equiv 0$ *has solution* $u = 1$ *of multiplicity* $\alpha/4$. *Function* $g_\alpha(u)$ *satisfies the following congruence (cf. with* (1.37))

$$g_\alpha(u) \equiv d_\alpha(u) g_\alpha(u^\alpha) (\text{mod } 2). \tag{1.39}$$

Proof. Formula (1.37) immediately follows from the definition (1.34) for $Q_\alpha(u)$ as an infinite product. Formulae (1.38) and (1.39) follow from (1.29). Note that

$$d_\alpha(u) \equiv \left(1+u^{-2+\alpha}\right)^{-1+\alpha/4} + u^{-1+2\alpha}\left(1+u^{-2+3\alpha}\right)^{-1+\alpha/4},$$

and $d_\alpha(u)$ in (1.34) has an even number of monomials with coefficient 1, and the number of monomials of even degree is equal to the number of monomials of odd degree. From here we derive that the equation $d_\alpha(u) \equiv 0$ has solution $u = 1$ of multiplicity $\alpha/4$.

Lemma (*Analyticity*) *For any* α, *the function* $Q_\alpha(u)$ *defined as an infinite product* (1.34) *is holomorphic in the open domain* $\Phi = \{u : |u| < 1\} \in \mathbb{C}$.

Proof. The product $\prod_{s=0}^{\infty}\left(1+u^{\alpha^s}\right)$ is holomorphic in the domain Φ because it is of exponential type [63].

Lemma (*Recurrence relations*) *Let the number* α be of the form 2^{x+2} for fixed $x = 0, 1, \ldots$. Then:

(a) *The following recurrence for the members of sequence* $\{q_k(\alpha)\}$ *from* (1.35) *holds*:

$$q_k(\alpha) = \sum_{t\in\Omega(k)} q_t(\alpha), \quad k = 1, 2, \ldots, \tag{1.40}$$

where the finite set $\Omega(k) = \{t = 0, 1, \ldots \mid s = 0, 1, \ldots, -1+\frac{\alpha}{4}$, *and tuple* (s,t) *satisfies one of the equations* $s(-2+\alpha)+\alpha t = k$, *or* $-1+2\alpha+s(-2+3\alpha)+\alpha t = k\}$. *Here, from the Lemma of Characteristic property,* $q_k(\alpha) = 1$ *if only a single summand in* (1.40) *is equal to 1, and all the others are equal to 0. Analogously,* $q_k(\alpha) = 0$ *iff all summands in* (1.40) *are equal to* 0.

(b) *Consider* $\alpha^{q-1} \le k < \alpha^q$, $k \in \mathbb{N}$ *and numbers*

$$F_\alpha(q,k) = \mathbf{res}_u u^{-k-1} \prod_{t=0}^{q-1} d_\alpha(u^{\alpha^t}). \tag{1.41}$$

Then the numbers $F_\alpha(q,k)$ *satisfy the following recurrence* [27]:

$$F_\alpha(q,k) = F_\alpha(q-1,k) + F_\alpha(q-1, k-\alpha^q+2\alpha^{q-1}). \tag{1.42}$$

In particular, since $q_k(\alpha) = F_\alpha(q,k)$, *then*

$$q_{-n+\alpha^{\sigma(n)}}(\alpha) = F_\alpha(\sigma(n)-1, -n+\alpha^{\sigma(n)}) + F_\alpha(\sigma(n)-1, -n+2\alpha^{\sigma(n)-1}).$$

Proof. Formula (1.40) follows from formula (1.34) for the series (1.35) and from the linearity rule for the **res** operator. Formula (1.42) follows from the definition of $F_\alpha(q,k)$ in (1.41) if under **res** sign in (1.41) we replace the last multiplier

$$d_\alpha\left(u^{\alpha^{r-1}}\right) = \sum_{s=0}^{-1+\alpha/4}\left(u^{s(-2+\alpha)\alpha^{r-1}} + u^{(-1+2\alpha+s(-2+3\alpha))\alpha^{r-1}}\right)$$

by the first term of the sum (corresponding to $s = 0$), that is equal to $1 + u^{(-2+\alpha)\alpha^{r-1}}$. This is possible since the value of the **res** operator for all other summands is obviously equal to zero.

Numerous equivalent formulations of the Collatz conjecture and its generalizations are given in [112]. It was noted in [28] that (1.33) is equivalent to the known number-theoretic reformulation of the problem [112]. In conclusion we give a new function-theoretic formulation of the Collatz conjecture, which directly follows from Theorem 2, the foregoing Lemmas and the definition of Collatz sequence.

Theorem 3. *Let the series $Q_\alpha(u)$ be defined by the infinite product* (1.34). *Then $3x+1$ conjecture is true iff for every natural n there are natural numbers $\sigma(n)$ and $\alpha = 2^{x+2}$, $x \in \mathbb{N}$, such that $n \le \alpha/4 - 1$ and one of the following equivalent conditions hold*:

(a) *The equality $q_{-n+\alpha^{\sigma(n)}}(\alpha) = 1$ holds.*

(b) *For the series $(d_\alpha(u))^{-1/(\alpha-1)} \in H(\mathbb{Q})$ the congruence $(d_\alpha(u))^{-1/(\alpha-1)} \equiv 1 (\mathrm{mod}\ 2)$ holds.*

(c) *$\exists$ function $Q_\alpha(u)$ of the form* (1.34) *analytic in the domain $\Phi = \{u : |u| < 1\} \in \mathbb{C}$, satisfying an integral equation of the form*

$$\frac{1}{2\pi i}\int_{\Gamma_\rho} Q_\alpha(u)\,\lambda(u,w)\,u^{-n+1}du = \tau_{\sigma(n)}(w), \tag{1.43}$$

where $\rho < 1$ and the integral is taken over the polydisc $\Gamma_\rho = \{u \in C : |u| = \rho\}$. Here

$$\tau_{\sigma(n)}(w) = \sum_{t=0}^{\infty} w^{\alpha^{2t+\sigma(n)}},\ \lambda(u,w) = w/u + \sum_{t=1}^{\infty} w^{\alpha^t}/u^{\alpha^t},$$

are holomorphic functions in the domain $\Psi_{\rho_1} = \{w : |w| \le \rho_1\} \in \mathbb{C}$, when $u \in \Gamma_\rho$ and $\rho_1 < \rho$.

In Theorem 2 we have found that the characteristic function $Q(u)$ of the stopping height in the Collatz conjecture is a Dirichlet series (exponential series). It is well know that the knowledge of the generating function means a lot in combinatorics. We have found several new properties of $Q(u)$ in [33]. However, the answer to the main question in the Collatz conjecture remains to be found. Our approach to this problem is to use the method of coefficients for exponential series. Note, that construction of the method of coefficients for these series requires not only writing inference rules and proving the Completeness Lemma, which is already done by the author for the Dirichlet series of usual type in section 1.2.3. In my opinion, the main work is to apply the method of coefficients to the hundreds of sums by divisors and p-adic expansions (well-known and new; see, for example, [8, 14, 16, 18, 21, 23, 104], etc.), as it was done in my book [25] for formal Laurent power series.

1.3.2 Computation of combinatorial sums in the theory of integral representations in $\mathbb{C}^n$

V. Krivokolesko and A. Tsikh (2005) discovered the following formulae of integral representations for functions holomorphic in linearly-convex polyhedrons in $\mathbb{C}^n$.

Theorem [57]. *Let $G = \{z : g^l(z,\overline{z}) < 0,\ l = 1,\dots,N\}$ be a bounded piecewise regular linearly convex domain in $\mathbb{C}^n$. Then every function $f(z)$ holomorphic in G and continuous in $\overline{G}$, is representable in G as*

$$f(z) = \sum_{k=1}^{n} (-1)^{k-1} \sum_{\sharp J=k} \sum_{|I|=n-k} \frac{I!}{(2\pi i)^n} \int_{S^J} \frac{f(\zeta) L_I\left(g^{j_1},\dots,g^{j_k}\right)}{\prod_{t=1}^{k} \langle \nabla g^{j_t}, \zeta - z\rangle^{i_t+1}} \omega_J, \qquad (1.44)$$

where $\sum_{\sharp J=k}$ stands for summation over ordered multi-indexes J of length $k : 1 \le j_1 < \ldots < j_k \le N$; $\sum_{|I|=n-k}$ stands for summation over ordered multi-indexes $I = (i_1,\dots,i_k)$ with the property $|I| := i_1 + \ldots + i_k = n-k$; L_I is the mixed Levian of order I, and $I! := i_1!\ldots i_k!$.

Corollary.

(a) *If $G = \{z : a_{j_l 1}|z_1| + \ldots + a_{j_l n}|z_n| - r_{j_l} < 0,\ l = 1,\dots,N\}$ and $k = 1$, then formula* (1.44) *breaks up into the sum of terms ν^j of the following type*:

$$\nu^j = \frac{(n-1)!(-1)^{n+p-1} r_j}{(2\pi i)^n a_{jp}} \cdot \prod_{m=1}^{n} a_{jm} \int_{|S^j|} d|\zeta|[p] \int_{|\xi|=1} \frac{f\left(\frac{|\zeta_1|}{\xi_1},\dots,\frac{|\zeta_n|}{\xi_n}\right)}{\left(r_j - \sum_{m=1}^{n} a_{jm} z_m \xi_m\right)^n} \cdot \frac{d\xi}{\xi},$$

where the sides $S^j = \{\zeta \in \overline{G} : a_{j1}|\zeta_1| + \ldots + a_{jn}|\zeta_n| = r_j\}$, $j = 1,\dots,N$.

(b) *If the edge $S^{j_1,\dots,j_n} = \{\zeta \in \overline{G} : a_{j_1 1}|\zeta_1| + \ldots + a_{j_1 n}|\zeta_n| = r_{j_1},\dots, a_{j_n 1}|\zeta_1| + \ldots + a_{j_n n}|\zeta_n| = r_{j_n}\}$ and $k = n$, then the formula* (1.44) *breaks up into the sum of no more than $\binom{N}{n}$ terms $\nu^{j_1,\dots,j_n}$ of the following type*:

$$\nu^{j_1,\dots,j_n} = \frac{(-1)^n |\zeta_1|\cdot\ldots\cdot|\zeta_n|}{(2\pi i)^n} \begin{vmatrix} a_{j_1 1} & \ldots & a_{j_1 n} \\ \ldots & \ldots & \ldots \\ a_{j_n 1} & \ldots & a_{j_n n} \end{vmatrix} \int_{|\xi|=1} \frac{f\left(\frac{|\zeta_1|}{\xi_1},\dots,\frac{|\zeta_n|}{\xi_n}\right)}{\prod_{t=1}^{n} \left(r_{j_t} - \sum_{m=1}^{n} a_{j_t m} z_m \xi_m\right)^n} \cdot \frac{d\xi}{\xi},$$

where $d|\zeta|[p] := d|\zeta_1|\cdot\ldots\cdot d|\zeta_{p-1}|\cdot d|\zeta_{p+1}|\cdot\ldots\cdot d|\zeta_n|$, $\xi := \xi_1\cdot\ldots\cdot\xi_n$, $\frac{d\xi}{\xi} = \frac{d\xi_1}{\xi_1} \wedge \ldots \wedge \frac{d\xi_n}{\xi_n}$.

The formula (1.44) was applied by V. Krivokolesko (2008) to polyhedrons of special type in n-circular domains in $\mathbb{C}^n$ [58]. On Reinchard's diagram these polyhedrons are convex polytopes. As a result the formula (1.44) essentially becomes simpler and an integration on border of domain is reduced to topological product of a unit polydisk and a projection of the boundary of the domain onto Reinchard's diagram.

In [58] various important partial cases of the integral formulae (1.44) were considered, which gives several interesting relations (the combinatorial identities) of type (1.45) and (1.46) between parameters of these integral representations:

$$\frac{(s_1+s_2+1)!}{(s_1)!(s_2)!}\sum_{m=0}^{s_2}\frac{(-1)^m}{s_1+m+1}\binom{s_2}{m}((1-\beta)^{s_1+m+1}-\alpha^{s_1+m+1})$$

$$=\sum_{k=0}^{s_2}\binom{s_2+k}{k}\left((1-\beta)^{s_1+1}\beta^k-\alpha^{s_1+1}(1-\alpha)^k\right),\ \forall s_1,s_2\geq 0, \qquad (1.45)$$

where $0<\alpha<1, 0<\beta<1$.

The identity (1.45) is obtained by integration of holomorphic monomials $z_1^{s_1}z_2^{s_2}$ on boundary $G=\{(z_1,z_2): |z_1|>1, |z_2|>1, \frac{|z_1|-1}{a}+\frac{|z_2|-1}{b}<1\}\subset\mathbb{C}^2$ and $\alpha=\frac{b}{a+b+ab}, \beta=\frac{b}{a+b+ab}$. By integration of holomorphic monomials $z_1^{s_1}z_2^{s_2}z_3^{s_3}$ on boundary of a linearly convex polyhedron $G=\{(z_1,z_2,z_3): |z_1|>1, |z_2|>1, |z_3|>1, a_{41}|z_1|+a_{42}z_2+a_{43}|z_3|-r_4<0\}\subset\mathbb{C}^3$ V. Krivokolesko obtained a series of identities of the following type:

$$\sum_J\Bigg\{(1-\alpha_{j_1}-\alpha_{j_2})^{1+s_{j_3}}\sum_{k=0}^{s_{j_2}}\sum_{l=0}^{s_{j_3}}\frac{(k+l+1+s_{j_1})!}{k!l!s_{j_1}!}a_{j_2}^k a_{j_3}^l$$
$$-\frac{(1+s_{j_1}+s_{j_2})!}{s_{j_1}!s_{j_2}!}\sum_{m=0}^{s_{j_2}}\frac{(-1)^m}{1+m+s_{j_1}}\binom{s_{j_2}}{m}\left\{\left(1-\frac{\alpha_{j_2}}{1-\alpha_{j_3}}\right)^{1+m+s_{j_1}}-\left(\frac{\alpha_{j_1}}{1-\alpha_{j_3}}\right)^{1+m+s_{j_1}}\right\}$$
$$\times(1-\alpha_{j_3})^{2+s_{j_1}+s_{j_2}}\sum_{k=0}^{s_{j_3}}\binom{1+k+s_{j_1}+s_{j_2}}{k}a_{j_3}^k\Bigg\}$$
$$+\frac{(2+s_1+s_2+s_3)!}{s_1!s_2!s_3!}\int_{\alpha_{j_1}}^{1-\alpha_{j_2}-\alpha_{j_3}}\int_{\alpha_{j_2}}^{1-\alpha_{j_3}-x}x^{s_1}y^{s_2}(1-x-y)^{s_3}\,dx\wedge dy=1,$$
$$\forall s_1,s_2,s_3=0,1,2,\ldots, \qquad (1.46)$$

where $\alpha_i:=a_{4i}/r_4, i=1,2,3$ *and* $\sum_J$ *stands for summation over all* $(2,1)$*-partitions of the* 3*-set* $\{1,2,3\}$. However the proof of these formulas demanded from the author various combinatorial and geometrical constructions and cumbersome calculations of determinants and integrals [58]. V. Krivokolesko and A. Tsikh have raised the question of an independent check of these identities. In the following Lemma we check the validity of (1.45) (validation of the formula (1.46) can be done in a similar manner, but because of bulkiness of standard calculations will be published by us in other work).

Lemma. *The formula* (1.45) *is valid.*

Proof. The standard scheme of the check of combinatorial identities consists usually in finding the generating functions in three variables (on the number of free parameters s_1,s_2,s_3) from left and right hand sides of these identities. However the structure of the sum in the left hand side of (1.45) allows us to prove it with the help of direct calculation. Let us denote by T the expression in the left hand side of (1.45). As

$$\frac{(s_1+s_2+1)!}{(s_1)!(s_2)!}\frac{1}{s_1+m+1}\binom{s_2}{m}=\binom{s_1+s_2+1}{s_2-m}\binom{s_1+m}{m},$$

then we get successively

$$T := \frac{(s_1+s_2+1)!}{(s_1)!(s_2)!}\sum_{m=0}^{s_2}\frac{(-1)^m}{s_1+m+1}\binom{s_2}{m}((1-\beta)^{s_1+m+1}-\alpha^{s_1+m+1})$$

$$= \sum_{m=0}^{s_2}(-1)^m\binom{s_1+s_2+1}{s_2-m}\binom{s_1+m}{m}((1-\beta)^{s_1+m+1}-\alpha^{s_1+m+1}) \qquad (1.47)$$

$$= S(1-\beta)-S(\alpha),$$

where

$$S(\alpha) := \alpha^{s_1+1}\sum_{m=0}^{s_2}(-\alpha)^m\binom{s_1+s_2+1}{s_2-m}\binom{s_1+m}{m}. \qquad (1.48)$$

Replacing the binomial coefficients in (1.48) according to the formula (1.2)

$$\binom{s_1+s_2+1}{s_2-m} = \mathbf{res}_x\frac{(1-x)^{-s_1-m-2}}{x^{s_2-m+1}}, \quad \binom{s_1+m}{m} = \mathbf{res}_y\frac{(1-y)^{-s_1-1}}{y^{m+1}},$$

we get

$$S(\alpha) = \alpha^{s_1+1}\sum_{m=0}^{s_2}(-\alpha)^m\,\mathbf{res}_x\frac{(1-x)^{-s_1-m-2}}{x^{s_2-m+1}}\times\mathbf{res}_y\frac{(1-y)^{-s_1-1}}{y^{m+1}} = \sum_{m=0}^{\infty}\cdots$$

(the interchange of the order of $\sum$ and $res_{x,y}$, and the separation of summands over the index m)

$$= \alpha^{s_1+1}\mathbf{res}_x\left\{\frac{(1-x)^{-s_1-2}}{x^{s_2+1}}\times\left[\mathbf{res}_y\sum_{m=0}^{\infty}\left(-\frac{\alpha x}{1-x}\right)^m\frac{(1-y)^{-s_1-1}}{y^{m+1}}\right]\right\}$$

(the summation over m in square brackets, the substitution rule and the substitution $y=-\alpha x/(1-x)\in H_1$)

$$= \alpha^{s_1+1}\mathbf{res}_x\left\{\frac{(1-x)^{-s_1-2}}{x^{s_2+1}}\times(1+\frac{\alpha x}{1-x})^{-s_1-1}\right\}$$

$$= \alpha^{s_1+1}\mathbf{res}_x\left\{\frac{(1-x)^{-1}(1-x(1-\alpha))^{-s_1-1}}{x^{s_2+1}}\right\}$$

$$= \alpha^{s_1+1}\mathbf{res}_x\left\{\frac{(\sum_{k=0}^{\infty}x^k)(\sum_{k=0}^{\infty}\binom{s_1+k-1}{k}(1-\alpha)^k x^k)}{x^{s_2+1}}\right\}$$

$$= \alpha^{s_1+1}\sum_{k=0}^{s_2}\binom{s_1+k-1}{k}(1-\alpha)^k,$$

i.e.

$$S(\alpha) = \alpha^{s_1+1}\sum_{k=0}^{s_2}\binom{s_1+k-1}{k}(1-\alpha)^k. \qquad (1.49)$$

Comparing the right and left hand sides of (1.45) using (1.47) – (1.49) we prove (1.45).

1.3.3 Combinatorial computations related to the inversion of a system of two power series in $\mathbb{C}^n$

Solving the problem of the inversion of the system of two power series V. Stepanenko ([102], 2008) has found out that it is equivalent to the problem of representation of the group $GL(2)$ in linear spaces of dimensions $(m+1)$, $m,\dots,2$ of homogeneous polynomials of various (arbitrarily large) degrees m. Let $m\in\mathbb{N}$, $p=1,\dots,m$, $q=1,\dots,\lfloor m/2\rfloor$, $p\geq q$, and $M=(M_{pq})$ be an $m\times(m+1)$ matrix of generators (the matrix of bases) of these spaces. He has also stated several interesting combinatorial problems of summation, connected with the study of the structure of the matrix M. For example, the sum of coefficients of the monomials of the polynomial M_{pq} in $(p+1)\times(p-q+1)$ variables is equal to the following expression:

$$S=S_{m,p,q}=(m-q)!q!\sum_A\left(\prod_{i=0}^{p-q}\prod_{j=0}^{q}(i!j!)^{-\alpha_{ij}}\frac{1}{\alpha_{ij}!}\right), \tag{1.50}$$

where the summation $\sum_A$ extends over integer nonnegative entries the of $(p+1)\times(p-q+1)$ matrix $A=(\alpha_{ij})$, which satisfy the following system of linear equations ($\alpha_{00}\equiv 0$):

$$\begin{aligned}\Sigma_1 &:= (1\alpha_{10}+0\alpha_{01})+\dots+(p\alpha_{p0}+(p-1)\,\alpha_{p-1,1}+\dots+(p-q)\,\alpha_{p-q,q})=m-q,\\ \Sigma_2 &:= (0\alpha_{10}+1\alpha_{01})+\dots+\ (0\alpha_{p0}+1\alpha_{p-1,1}+\dots+q\alpha_{p-q,q})=q,\\ \Sigma_3 &:= (\alpha_{10}+\alpha_{01})+\dots+\ (\alpha_{p0}+\alpha_{p-1,1}+\dots+\alpha_{p-q,q})=m-p+1.\end{aligned} \tag{1.51}$$

V. Stepanenko (2008) has stated the following problem: *calculate (if it is possible) the sum $S_{m,p,q}$ in* (1.50) *in integer parameters m, p, q ($p\geq q$) with linear restrictions* (1.51) *on* $(p+1)\times(p-q+1)$ *summation indexes α_{ij} in closed form.*

Lemma. *The following formula is valid*

$$S_{m,p,q}=S_2(m-p+1,m), \tag{1.52}$$

where $S_2(n,m)$ are Stirling numbers of the second kind.[2]

Proof. Replacing the exponential coefficients in (1.50) by the formula (1.3)

$$(i!j!)^{\alpha_{ij}}/\alpha_{ij}!=\mathbf{res}_{t_{ij}}(\exp(t_{ij}/i!j!)\,t_{ij}^{-\alpha_{ij}-1}),\forall i,j,$$

we get

$$S=(m-q)!q!\sum_A\left\{\prod_{\forall i,j}i!j!^{-\alpha_{ij}}\frac{1}{\alpha_{ij}!}\right\}$$

[2] Observe, that the right-hand side of formula (1.52) does not depend on the parameter q.

$$= (m-q)!q! \sum_{\forall \alpha_{ij}=0,1,\dots} \left\{ \prod_{\forall i,j} \mathbf{res}_{t_{ij}} (\exp(t_{ij}/i!j!)\, t_{ij}^{-\alpha_{ij}-1}) \right.$$

$$\left. \times \delta\left(m-q, \textstyle\sum_1\right) \times \delta\left(q, \textstyle\sum_2\right) \times \delta\left(m-p+1, \textstyle\sum_3\right) \right\}. \qquad (1.53)$$

Last three factors are added here for the account in the sum (1.50) of each of three linear restrictions (1.51) on the set of summation indexes α_{ij}, that has allowed us to distribute summation in (1.53) on all values $\alpha_{ij} = 0, 1, \dots$(see, [25], § 5.2; [61]). Thus by formula (1.4) for $\delta(n,k)$ we obtain

$$S = (m-q)!q! \sum_{i=0}^{p} \sum_{j=1}^{p-q} \sum_{\alpha_{ij}=0}^{\infty} \left\{ \prod_{\forall i,j} \mathbf{res}_{t_{ij}} (\exp(t_{ij}/i!j!)\, t_{ij}^{-\alpha_{ij}-1}) \right.$$

$$\left. \times \mathbf{res}_x x^{-m+q+\Sigma_1-1} \times \mathbf{res}_y y^{-q+\Sigma_2-1} \times \mathbf{res}_z z^{-m+p-2+\Sigma_3} \right\}$$

$$= (m-q)!q! \sum_{i=0}^{p} \sum_{j=1}^{p-q} \sum_{\alpha_{ij}=0}^{\infty} \mathbf{res}_{x,y,z} \left\{ \prod_{\forall i,j} \mathbf{res}_{t_{ij}} \left(\exp(t_{ij}/i!j!)\, t_{ij}^{-\alpha_{ij}-1} \right.\right.$$

$$\left.\left. \times x^{-m+q+\Sigma_1-1}\, y^{-q+\Sigma_2-1}\, z^{-m+p-2+\Sigma_3} \right) \right\}$$

(interchanging the order of sums $\sum_{i,j,\alpha_{ij}}$ and the operator $\mathbf{res}_{x,y,z}$)

$$= (m-q)!q! \mathbf{res}_{x,y,z} \{ x^{-m+q+\Sigma_1-1}\, y^{-q+\Sigma_2-1}\, z^{-m+p-2+\Sigma_3}$$

$$\times \sum_{i=0}^{p} \sum_{j=1}^{p-q} \sum_{\alpha_{ij}=0}^{\infty} \prod_{\forall i,j} \mathbf{res}_{t_{ij}} (\exp(t_{ij}/i!j!)\, t_{ij}^{-\alpha_{ij}-1}) \}$$

(separating in the last expression factors with degrees of variables x, y and z, which are contained in the sums Σ_1, Σ_2 and Σ_3)

$$= (m-q)!q! \mathbf{res}_{x,y,z} \{ x^{-m+q-1}\, y^{-q-1}\, z^{-m+p-2} \times$$

$$\prod_{\forall i,j} [\sum_{\alpha_{ij}=0}^{\infty} (x^i y^j z)^{\alpha_{ij}}\, \mathbf{res}_{t_{ij}} [\exp(t_{ij}/i!j!)\, t_{ij}^{-\alpha_{ij}-1}]] \}$$

(summing over each index α_{ij} in square brackets: the substitution rule for each variable t_{ij}, and the substitutions $t_{ij} = x^i y^j z$, $\forall i, j$)

$$= (m-q)!q! \mathbf{res}_{x,y,z} \{ x^{-m+q-1}\, y^{-q-1}\, z^{-m+p-2} \{ \textstyle\prod_{\forall i,j} \exp\left(x^i y^j z/i!j!\right) \} \}$$

$$= (m-q)!q! \mathbf{res}_{x,y,z} \{ x^{-m+q-1}\, y^{-q-1}\, z^{-m+p-2} \exp(-z + \sum_{i=0}^{p} \sum_{j=0}^{p-q} x^i y^j z/i!j!) \}$$

(by definition of the operator $\mathbf{res}_z$)

$$= \frac{(m-q)!q!}{(m-p+1)!}\mathbf{res}_{x,y}\{x^{-m+q-1}y^{-q-1}(-1+\sum_{i=0}^{p}\sum_{j=0}^{p-q}x^iy^j/i!j!)^{m-p+1})\}$$

(as $p \geq q$, then by definition of the operator $\mathbf{res}_{x,y}$)

$$= \frac{(m-q)!q!}{(m-p+1)!}\mathbf{res}_{x,y}\{x^{-m+q-1}y^{-q-1}(-1+\sum_{i=0}^{\infty}\sum_{j=0}^{\infty}x^iy^j/i!j!)^{m-p+1})\}$$

(by the formula $\sum_{\forall i,j} x^iy^j/i!j! = \exp(x)\times\exp(x) = \exp(x+y)$)

$$S = \frac{(m-q)!q!}{(m-p+1)!}\mathbf{res}_{x,y}\{x^{-m+q-1}y^{-q-1}(-1+\exp(x+y))^{m-p+1})\}$$

(by the following substitution $x = yX$ and then $Y = y(1+X)$)

$$= \frac{(m-q)!q!}{(m-p+1)!}\mathbf{res}_{X,y}\{X^{-m+q-1}y^{-m-1}(-1+\exp(y(1+X)))^{m-p+1})$$

$$= \frac{(m-q)!q!}{(m-p+1)!}\mathbf{res}_{X,Y}\{X^{-m+q-1}Y^{-m-1}(1+X)^m(-1+\exp Y)^{m-p+1})$$

$$= \frac{(m-q)!q!}{(m-p+1)!}\mathbf{res}_X\{X^{-m+q-1}(1+X)^m\}\times\mathbf{res}_Y\{Y^{-m-1}(-1+\exp Y)^{m-p+1}\}$$

(by the formulas (1.1) and (1.3))

$$= \frac{(m-q)!q!}{(m-p+1)!}\times\binom{m}{m-q}\times\frac{(m-p+1)!}{m!}S_2(m-p+1,m) = S_2(m-p+1,m).$$

□

1.4 Algebraic characterization of the method of coefficients as a method of summation

Here we shall give a new algebraic characterization of the method of coefficients, which is based on the φ-operation of isomorphism, generated by the classical one-to-one mapping φ between the set $\mathscr{A}$ of numerical sequences and the set $\mathscr{B}$ of generating series of a given type.

In [25, 31] an extensive list of open problems connected with the method of coefficients for various types of generating series was presented. This method has been successfully explored by various authors [49, 53, 115] and has found many applications to concrete problems of summation [17, 45, 61, 62, 64, 73, 74, 92, 103] and others. We also mention the applications to computer algebra [31, 38, 41, 79, 80, 105] and to physics [67, 68]. Another example is the excellent results of Professor Ch. Krattenthaler, related to the use of the method of coefficients in the context of Euler and interpolation series in one and several variables and its many applications [53, 54, 55, 56]. The method of coefficients also has been extended in the papers by A. Yuzhakov, I-C. Huang and G. Xin ([2, 50, 51, 73, 114], and others), in connection with combinatorial applications of the theory of multidimensional residues in $\mathbb{C}^n$. The idea of calculation of a combinatorial sum by means of its integral representation has been further developed in the books [7, 19, 22, 37, 72, 76, 78, 98, 107, 115] and also in the remarkable papers [3, 10, 36, 40, 66, 96, 111], and others.

At the same time in many interesting combinatorial publications ([44, 75], etc.) the authors are usually restricting in their calculations and notations to the traditional application of the method of generating functions as a tool for deriving and proving combinatorial identities without proving completeness of the used calculus.

Consider the important concept of *the isotopy of operations* on a groupoid G [9]. Let α, β and γ be arbitrary one-to-one mappings of G onto itself. The binary operations $\times$ and $\otimes$ on G are called *isotopical*, if

$$x \otimes y = \gamma^{-1}(\alpha(x) \times \beta(y)),\ \forall\, x,\ y \in G. \tag{1.54}$$

Let $\overline{\mathbb{R}} = \mathbb{R} \cup (\infty)$, $D_1, D_2 \subset \overline{\mathbb{R}}$, and let $\varphi : D_1 \to D_2$ be an arbitrary one-to-one mapping. One of the problems listed in [26] uses the following well-known nonstandard isotopical operations $\oplus$ and $\otimes$ over number fields:

$$x \oplus y := \varphi\left(\varphi^{(-1)}(x) + \varphi^{(-1)}(y)\right),\ x \otimes y := \varphi\left(\varphi^{(-1)}(x) \times \varphi^{(-1)}(y)\right),\ x, y \in D_2. \tag{1.55}$$

From (1.55) we immediately obtain the following dual formulas:

$$x + y = \varphi\left(\varphi^{(-1)}(x) \oplus \varphi^{(-1)}(y)\right),\ x \times y = \varphi\left(\varphi^{(-1)}(x) \otimes \varphi^{(-1)}(y)\right),\ x, y \in D_1. \tag{1.56}$$

For example, in the special case $\varphi(x) = \ln(x)$, $x > 0, x \in \overline{\mathbb{R}}$, , $\varphi^{(-1)}(x) = \exp(x)$, $x \in \mathbb{R}$, we get

$$x \otimes y := \varphi\left(\varphi^{(-1)}(x) \times \varphi^{(-1)}(y)\right) = \ln(\exp(x) \times \exp(y))$$

$$= x + y,\ x, y \in \mathbb{R}, \tag{1.57}$$

$$x \oplus y := \varphi\left(\varphi^{(-1)}(x) + \varphi^{(-1)}(y)\right) = \ln(\exp(x) + \exp(y)),\ \text{if}\ x, y \in \mathbb{R}, \tag{1.58}$$

i.e., the operation of multiplication transforms into the operation of addition; if $\varphi(x) = 1/x$, $\varphi^{(-1)}(x) = 1/x$, $x \in \overline{\mathbb{R}}$, then the operation $r_1 \oplus r_2$ in (1.55) obviously generates the formula of resistance of an electric circuit with parallel connection of two conductors with resistances r_1 and r_2. Note ([5], p.11), that the tropical operation $x * y := \max(x, y)$ for real x and y obtained from the formula (1.58)

$$\frac{x *_h y}{h} := \ln(\exp(x/h) + \exp(y/h)), \tag{1.59}$$

as the quantum-mechanical "short-wave limiting transition" as the wave length h approaches zero. In other words, from (1.59) the formula of tropical operation $\max(x,y)$ follows for $h \to 0$. In [25] the φ-calculus over numerical fields has been extended to φ-calculus for matrices in several forms, that allowed us to obtain a number of new interesting isoperimetric inequalities for matrix functions [30].

Let $\mathbb{C}[[x]]$ denote the set of formal Laurent power series containing a finite number of terms with negative degrees, and let $\mathscr{A}$ be the set of numerical sequences $\{a_n\}$, and $A(x) = \sum_n a_n x^n \in \mathbb{C}[[x]]$ be the generating function of power type for the sequence $\{a_n\}$. For $A(x) \in \mathbb{C}[[x]]$ define once more the *formal residue* as

$$\mathbf{res}_x A(x) = a_{-1}.$$

Thus we have the pair of inverse transforms $\varphi : \mathscr{A} \to \mathbb{C}[[x]]$ and $\varphi^{(-1)} : \mathbb{C}[[x]] \to \mathscr{A}$ of the following type:

$$\varphi : A(x) = \sum\nolimits_n a_n x^n,\ \{a_n\} \in \mathscr{A};\quad \varphi^{(-1)} :\ a_n = \mathbf{res}_x A(x)\, x^{-n-1},\ \forall n,\ A(x) \in \mathbb{C}[[x]]. \tag{1.60}$$

Furthermore it follows directly from the definition (1.60) for $\{a_n\}$ that, for example, the rule of additivity for the **res** operator holds:

$$\mathbf{res}_x A(x)\, x^{-n-1} + \mathbf{res}_x B(x)\, x^{-n-1} = \mathbf{res}_x (A(x) + B(x)) x^{-n-1},\ \forall n. \tag{1.61}$$

This rule gives by induction the property of commutativity for the operators $\sum$ and **res** : *if* $A_1(x), A_2(x), \ldots \in \mathbb{C}[[x]]$, *then*

$$\sum\nolimits_k \mathbf{res}_x A_k(x)\, x^{-n-1} = \mathbf{res}_x \left(\sum\nolimits_k A_k(x)\right) x^{-n-1}. \tag{1.62}$$

Analogously the substitution rule for the **res** operator follows from the definition (1.60) for $\{a_n\}$:

$$A(x) := \sum\nolimits_n a_n x^n = \sum\nolimits_n x^n \mathbf{res}_x A_k(x)\, x^{-n-1},$$

i.e.,

$$\sum\nolimits_n x^n \mathbf{res}_z A(z)\, z^{-n-1} = [A(z)]_{z=x} = A(x). \tag{1.63}$$

Note that under the same scheme the elementary "school" identity is deduced

$$\exp(\ln(x)) = x,\ x > 0. \tag{1.64}$$

In fact, we obtain

$$x := \exp(y) \Rightarrow y := \ln(x),\ x > 0 \Rightarrow (1.64).$$

The concepts (1.54) and (1.55) allow one to formulate the new

Definition. *Let* φ *be the one-to-one mapping* $\varphi : \mathscr{A} \to \mathbb{C}[[x]]$ and $\varphi^{(-1)} : \mathbb{C}[[x]] \to \mathscr{A}$. *The inference rule of the method of coefficients is called the* φ*-operation of isomorphism if it can be interpreted as formula of type* (1.55) *or type* (1.56).

Lemma. *The linearity rule and other inference rules of the method of coefficients for formal Laurent power series are φ-operations of isomorphism.*

Proof. Rewrite these rules as pairs of inverse transforms (1.60). We shall denote the addition operation for the sequences $\{a_n\}$ and $\{b_n\}$ from $\mathscr{A}$ by $+$, and the addition operation for the series $A(x)$ and $B(x)$ from $C[[x]]$ by $\oplus$. Now we have from (1.60)

$$A(x) \oplus B(x) := \sum_n x^n (a_n + b_n)$$
$$= \sum_n x^n (\mathbf{res}_x A(x) x^{-n-1} + \mathbf{res}_x B(x) x^{-n-1})$$
$$:= \varphi\left(\varphi^{(-1)}(A(x)) + \varphi^{(-1)}(B(x))\right),$$

i.e.,

$$A(x) \oplus B(x) = \varphi\left(\varphi^{(-1)}(A(x)) + \varphi^{(-1)}(B(x))\right), \text{ if } A(x), B(x) \in \mathbb{C}[[x]]. \quad (1.65)$$

Conversely, we get

$$\{a_n\} + \{b_n\} = \{\mathbf{res}_x A(x) x^{-n-1}\} + \{\mathbf{res}_x B(x) x^{-n-1}\}$$
$$= \{\mathbf{res}_x (A(x) \oplus B(x)) x^{-n-1}\} := \varphi^{(-1)}(\varphi(\{a_n\}) \oplus \varphi(\{b_n\})),$$

i.e.,

$$\{a_n\} + \{b_n\} = \varphi^{(-1)}(\varphi(\{a_n\}) \oplus \varphi(\{b_n\})), \; if \; \{a_n\}, \{b_n\} \in \mathscr{A}. \quad (1.66)$$

It is equally easy to give an interpretation for the substitution rule (1.63)

$$\sum_n x^n \mathbf{res}_z A(z) z^{-n-1} = [A(z)]_{z=x} = A(x),$$

as a formula of type (1.55) for the substitution operation for a series in $\mathbb{C}[[x]]$.

Note also, that the identity (1.64) is the *φ-operation of isomorphism* at $\varphi(x) = \ln(x)$ which translates the multiplication operation into the addition operation and is successfully used, for example, in the transition from studying Lie groups to studying Lie algebras (Serre, Pontrjagin, and others). This identity is also directly used in algebraic and combinatorial calculations as an inference rule.

From the last Lemma we can deduce the following important

Conclusion. *In the method of coefficients as a calculus method, simultaneous use of the pair of direct and inverse φ-transforms* (1.60) *is directly incorporated in each inference formula.*

Remark. *Using the same scheme of calculations as above we can obtain a method of coefficients (the set of inference rules and the Completeness Lemma) and its algebraic characterization for several new classical types of generating series as it is done above for formal Dirichlet series (exponential series). This allows us to obtain a new and uniform proof of a number of well-known formulae for classical functions of number theory (see, for example,* [8], § 17; [106], *Chapter* 1), *as well as the congruences by* mod p (p -prime) *for several combinatorial sums, including* *the congruences for coefficients of the generating function of the stopping height in the Collatz conjecture* [27].

These results allow one to formulate the following statement [32].

E-principle of summation: *each pair of inverse linear transforms (for sequences, series, functions, etc.), independently of the way of definition of the one-to-one mapping* φ, *generates the corresponding method of summation (the method of coefficients).*

It is also easy to check, that a similar construction arises in standard calculations by means pairs of Mellin transforms, Fourier transforms, Laplace transforms, Radon transforms, G-transforms and many other classic linear integral transforms. A remarkable example of new applications of such transforms are the recent results of Krasnoyarsk mathematicians I. Antipova (2001, [4]) and V. Stepanenko (2003, [100]). V. Stepanenko has subsequently applied the direct and inverse Mellin transforms (1.69),(1.70) to each monomial

$$y^{\mu}(x) = y_1^{\mu_1}(x)\dots y_n^{\mu_n}(x) \tag{1.67}$$

of the solution $y = y(x) = (y_1(x), \dots, y_n(x))$ to the system of multivariate algebraic equations with complex coefficients (in the normal form)

$$y_i^{\mu_i}(x) + \sum_{k=1}^{p_i} x_{m^i_{1k}\dots m^i_{nk}} y_1^{m^i_{1k}} \dots y_n^{m^i_{nk}} - 1 = 0,\ i = 1,\dots,n. \tag{1.68}$$

Thus he obtained the following integral representation of this monomial:

$$y_1^{\mu_1}(x)\dots y_n^{\mu_n}(x) = \frac{1}{(2\pi i)^{|p|}} \int_{\gamma + i\mathbb{R}^{|p|}} M[y^{\mu}](u) \prod\nolimits_{i=1}^{n} \prod\nolimits_{s=1}^{p_i} \left(x_s^i\right)^{-u_s^i - 1} du, \tag{1.69}$$

where $\mu = (\mu_1, \dots, \mu_n) \in \mathbb{R}^n_+$, including limiting $\mu = e_i$ $(i = 1, ..., n)$ in $\mathbb{R}^n$, $|p| = p_1 + \dots + p_n$, $du = du_1^1 \wedge \dots \wedge du_{p_1}^1 \wedge \dots \wedge du_1^n \wedge \dots \wedge du_{p_n}^n$, and

$$M[y^{\mu}](u) = \int_{\mathbb{R}_+^{|p|}} \prod\nolimits_{i=1}^{n} \prod\nolimits_{s=1}^{p_i} \left(x_s^i\right)^{u_s^i - 1} dx, \tag{1.70}$$

where $dx = dx_1^1 \wedge \dots \wedge dx_{p_1}^1 \wedge \dots \wedge dx_1^n \wedge \dots \wedge dx_{p_n}^n$. Finally, V. Stepanenko used the method of separating cycles of A. Tsikh's and one of A. Marichev's formula to calculate multiple integrals over skeleton of a polydisc in $\mathbb{C}^n$. This allowed him to find all solutions of an arbitrary system of algebraic equations by means of multiple formal Laurent power series of hypergeometric type [100, 101].

The general theory of integral transforms contains many impressive results of the same type. For example, A. Plamenevskii [82] gives the description of the algebra of pseudo-differential operators of discontinuous symbols on manifolds in particular by using the integral operators of type $M^{-1}EM$, where E is some integral transform of functions on the $(n-1)$-dimensional sphere in $\mathbb{R}^n$. Here the isomorphism between corresponding classes of functions is obtained by the direct incorporation of the inverse pair of M-transforms in the integral formula for this transform (as well as for effective G-transform [93]).

The history of several fundamental mathematical problems shows, that in many cases the success of investigations is directly connected with the presence of a general combinatorial scheme of its solution. This scheme treats the tree of various variants of solutions and allocates crucial points in each of them (see, for example, [24, 108]). In our case, according to the E-principle the method of coefficients plays the role of general combinatorial scheme (the inference rules and the Completeness Lemma), and the successful overcoming of computational difficulties depends on the completeness of the list of expansions for analytical functions of desired type. In several examples from calculus it has been shown [25, 82, 100, 115], that the main role here is played by the pairs of inverse transforms of type (1.60), the tables of integrals and general theory of integrals of desired type. Among them are the well-known integral transforms of Cauchy, Mellin, Fourier, Laplace, etc. (see also Appendix), giving many interesting applications in computer algebra and combinatorics (see, for example, [1, 72] and others).

Combinatorics and other fields of mathematics put forward a multitude of summation tasks of various types (see [24, 69, 99, 108], and also [8, 43, 86, 89, 94, 97]). The E-principle provides a foundation for the classical method of generating functions (generating integrals) as a method of summation for different classes of generating series. It also makes possible to reduce the variety of calculations with them to a uniform combinatorial scheme, as well as to set up a new extensive program of open summation problems, which is based on the construction and regular search of pairs of direct and inverse transforms of various types (for sequences, functions, etc.). Now we can expand the list of open problems in [25, 31], connected with the solution of problems of summation and the solution of equations over various number fields and other algebraic systems, including the tropical calculus [12, 13], the umbral calculus [6, 52, 90, 109], φ-calculus [9, 20], the calculus over noncommutative algebraic systems [39, 99], and its applications. These problems are especially important for finite fields [9, 14, 77], including regular use of pairs of inverse discrete Fourier transform and Z-transform [11, 48, 83, 88, 110, 113]. It is interesting also for the algebraic characterization of calculus and corresponding functional equations which is based on the isotopical operations [9], generated by combinatorial mapping of various types [65]. Following L. Euler, G. Pólya [84], G.-C. Rota [24, 91] we promote the idea of unity of discrete and continuous mathematics in the field of summation problems in computer algebra [42, 81, 116].

Acknowledgements

I would like to thank my colleagues and friends M. Davletshin, M. Golovanov, I. Kotsireas, V. Krivokolesko, A. Machnev, V. Stepanenko, T. Sadykov, S. Tsarev and E. Zima for fruitful discussions, numerous comments and useful remarks.

Appendix. Table of pairs of classical integral transforms and their inference rules

1. *A locally integrable functions on* $(0,\infty)$ *is one that is absolutely integrable on all closed subintervals of* $(0,\infty)$. *The direct Mellin transform* of a locally integrable function $f(x)$ on $(0,\infty)$ is defined by

$$F(s) = M[f](s) = \int_0^\infty x^{s-1} f(x)\,dx, \tag{1.71}$$

when the integral converges. If $M[f](s)$ is analytic in the strip $a < \mathbf{Re}(s) < b$, then *the inverse transform of Mellin* (*the Mellin – Barns integrals*) is given by

$$f(x) = M^{-1}[F(s)(x)] = \frac{1}{2\pi i}\int_{c-i\infty}^{c+i\infty} x^{-s} F(s)\,ds,\ a < c < b, \tag{1.72}$$

which is valid at all points $x > 0$ where $f(x)$ is continuous. Then the following pair of integral representations holds:

$$\frac{1}{2\pi i}\int_{c-i\infty}^{c+i\infty} x^{-s}\left(\int_0^\infty t^{s-1} f(t)\,dt\right) ds = f(x), \tag{1.73}$$

$$\int_0^\infty x^{s-1}\left(\frac{1}{2\pi i}\int_{c-i\infty}^{c+i\infty} x^{-t} F(t)\,dt\right) dx = F(s),\ a < c < b. \tag{1.74}$$

For important special cases of Mellin transform see, for example, [72].
2. *The direct Fourier transform* and *the inverse Fourier transform* are related by the pair of formulae

$$F(x) = F[f(t);x] = \frac{1}{\sqrt{2\pi}}\int_{-\infty}^{+\infty} f(t)\,e^{itx}\,dt, \tag{1.75}$$

$$f(t) = F^{-1}[F(x);t] = F[F(x);-t] = \frac{1}{\sqrt{2\pi}}\int_{-\infty}^{+\infty} F(x)\,e^{-ixt}\,dx. \tag{1.76}$$

Then the following integral representations are valid:

$$f(x) = \frac{1}{\pi}\int_0^{+\infty}\left(\int_{-\infty}^{+\infty} f(t)\cos(u(x-t))\,dt\right) du \tag{1.77}$$

$$= \frac{1}{2\pi}\int_{-\infty}^{+\infty} e^{-ixu}\left(\int_{-\infty}^{+\infty} f(t)\,e^{-iut}\,dt\right) du. \tag{1.78}$$

The formula (1.77) is called the *Fourier's integral formula*. Independently of Fourier, Cauchy obtained the equivalent formula (1.78), called the *exponential form of Fourier's integral formula*.
3. *The direct Laplace* and *the inverse Laplace transform* are related by the pair of formulae

$$g(p) = \int_0^{+\infty} f(t)\, e^{ipt}\, dt, \; f(t) = \frac{1}{2\pi i} \int_{c-i\infty}^{c+i\infty} e^{zt} g(z)\, dz, \tag{1.79}$$

which generates the following pair of integral representations:

$$\begin{aligned} \frac{1}{2\pi i} \int_{c-i\infty}^{c+i\infty} e^{zt} \left(\int_0^{+\infty} f(t)\, e^{izt} dt \right) dz = f(t)\,, \\ \frac{1}{2\pi i} \int_0^{\infty} e^{ipt} \left(\int_{c-i\infty}^{c+i\infty} e^{zt} g(z)\, dz \right) dt = g(p). \end{aligned} \tag{1.80}$$

References

1. *Abramov S.A. and Tsarev S.P.* (1997). Peripheral factorization of linear ordinary operators, Programming & Computer Software, No. 1, p. 59–67.
2. *Aizenberg L.A. and Yuzhakov A.P.* (1979). Integral representation and residues in multidimensional complex analysis. Nauka, Novosibirsk (in Russian).
3. *Andrews G.E.* (1970). On the foundations of combinatorial theory. IV. Finite vector space and Eulerian generating functions. Stud. Appl. Math. **49**, 239–258.
4. *Antipova I.A.* (2001). Mellin transforms for superposition of the general algebraic functions. Proc. Intern. Conf. "Mathematical models and methods of their investigations", vol. **1**, Krasn. State Univ., Krasnoyarsk, 31–35 (in Russian).
5. *Arnold V.I.* (2005). Dynamics, statistic and projective geometry of Galois fields. Publ. MCCME, M. (in Russian).
6. *Barnabei M., Brini A. and Nicoletti G.* (1982). Recursive matrices and umbral calculus. J. Algebra **75**, 546–573.
7. *Balser W.* (1994). From divergent power series to analytic functions. Theory and application of multisummable power series. Lecture Notes in Mathematics, **1582**, Springer-Verlag, Berlin.
8. *Bateman G. and Erdélyi A.* (1955). Higher transcendental functions, vol. **3:** Chapter 19. Mc Graw-Hill Comp., New York.
9. *Belousov V.D.* (1967). Foundations of the theory of quasigroups and loops. Nauka, Moscow, 223 pages (in Russian).
10. *Bertozzi A. and McKenna J.* (1993). Multidimensional residues, generating functions, and their application to queueing netwoks. SIAM Review **35**: 2, 239–268.
11. *Campello de Souza R.M., de Oliveira H.M. and Silva D.* (2002). The Z transforms over Finite Fields. Intern. Telecom. Symp. – ITS2002, Natal, Brasil, 6 pages.
12. *Cao Z.Q., Kim K.H. and Roush F.W.* (1984). Incline Algebra and Applications. John Wiley, New York.
13. *Cuninghame-Green R.A.* (1979). Minimax Algebra. Lect. Notes in Economics and Mathematical Systems **166**, Springer, Berlin.
14. *Carlitz L.* (1932). The arithmetic of polynomials in a Galois Field. Amer. J. Math. **54**, 39–50.
15. *Cartan H.* (1961). Théorie élémentaire des fonctions analytiques d'une on plusieurs variables complexes. Hermann, Paris.
16. *Chamberland M. and Dilcher K.* (2006). Divisibility properties of a class of binomial sums. J. Number Theory **120**, 349–371.
17. *Chen W.Y.C., Qin J., Reidys C.M. and Zeilberger D.* (2008). Efficient counting and asymptotic of k-noncrossing tangled-diagrams. Electron. J. Combin. **16** (2009), no. 1, Research Paper 37.
18. *Cheng S.E.* (2003). Generating function proofs of identities and congruences. PhD thesis, Michigan State Univ., Michigan, 86 pages.

19. *Consul P.C. and Famoye F.* (2006). Lagrangian probability distributions. Birkhäuser Boston Inc., Boston, MA.
20. *Cooke D.J. and Bez H.E.* (1984). Computer mathematics. Cambridge Univ. Press, Cambridge.
21. *Dickson L.E.* (1966). History of the Theory of Numbers, vol. **1**, Chelsea Publishing Co., New York.
22. *Dingle R.B.* (1973). Asymptotic expansions: their derivation and interpretation. Acad. Press, New York.
23. *Deng Y.* (2006). A class of combinatorial identities. Discrete Math. **306**, 2234–2240.
24. *Doubilet P., Rota G.-C. and Stanley R.* (1972). On the foundations of combinatorial theory. VI: The idea of generating function. In: Proc. Sixth Berkeley Sympos. on Math. Stat. and Prob. (1970/71): vol. **II**. Prob. Theory, Univ. California Press, Berkeley, CA, 267–318.
25. *Egorychev G.P.* (1977). Integral representation and the computation of combinatorial sums. Novosibirsk, Nauka (in Russian); English: Transl. of Math. Monographs **59**, AMS, 1984, 2-nd Ed. in 1989.
26. *Egorychev G.P.* (2000). Algorithms of integral representation of combinatorial sums and their applications. Proc. of 12-th Intern. Conf. on Formal Power Series and Algebraic Combinatorics (FPSAC 2000), Moscow, Russia, June 2000, 15–29.
27. *Egorychev G.P.* (2004). Solution of the Margenstein-Matiyasevich's question in $3x+1$ problem. Preprint ISBN 5-7636-0632-9, Krasnoyarsk State Technical Univ., Krasnoyarsk, 12 pages (in Russian).
28. *Egorychev G.P. and Zima E.V.* (2004). The characteristic function in $3x+1$ problem. Proc. Intern. School-Seminare "Synthesis and Complexity of Management Systems", Math. Inst. of Sib. Branch of Russian Acad. Nauk, Novosibirsk, 34–40 (in Russian).
29. *Egorychev G.P. and Zima E.V.* (2005). Decomposition and group theoretic characterization of pairs of inverse relations of the Riordan type. Acta Appl. Math. **85**, 93–109.
30. *Egorychev G.P.* (2008). Discrete Mathematics. Permanents. Sib. Federal Univ., Krasnoyarsk, 272 pages (in Russian).
31. *Egorychev G.P. and Zima E.V.* (2008). Integral representation and algorithms for closed form summation. Handbook of Algebra, vol.5, (ed. M. Hazewinkel), Elsevier, 459–529.
32. *Egorychev G.P.* (2008). Method of coefficients: an algebraic characterization and recent applications. Issues of VII Intern. School-conf. of Theory Group (Cheljabinsk, Russia, August 3–9, 2008), Inst. Math. and Mech. Ural. Otdel. RAN, Ekaterinburg, 2 pages.
33. *Egorychev G.P. and Zima E.V.* (2008). Collatz conjecture from the integral representation point of view. Inst. Math. and Mech. Ural. Otdel. RAN, 12 pages (to appear).
34. *Evgrafov M.A.* (1968). Analytical functions. Nauka, M.
35. *Evgrafov M.A.* (1986). Series and integral representations. Itogi Nauki i Techniki. Sovr. Problems Mat., Fund. Napr. **13**, VINITI, M., 5–92 (in Russian).
36. *Flajolet P. and Salvy B.* (1998). Euler sums and contour integral representations. Experiment. Math. **7**, 15–35.
37. *Flajolet P. and Sedgewick R.* (2007). Analytic Combinatorics. Cambridge University Press, Cambridge, 2009.
38. *Gerhard J., Giesbrecht M., Storjohann A. and Zima E.* (2003). Shiftless decomposition and polynomial-time rational summation. *Proc. of* ISSAC 2003, ACM Press, 119–126.
39. *Gessel I.M.* (1980). A noncommutative generalization and q-analog of the Lagrange inversion formula. Trans. Amer. Math. Soc. **257**, 455–482.
40. *Gessel I.M.* (1997). Generating functions and generalized Dedekind sums. Elec. J. Comb., 4, Wilf Festschrift, R11.
41. *Greene D.H. and Knuth D.E.* (1981). Mathematics for the analysis of algorithms. Birkhäuser, Boston.
42. *Gosper R.W.* (1978). Decision procedure for indefinite hypergeometric summation. Proc. Natl. Acad. Sci. USA **75**, 40–42.
43. *Gould H.W.* (1972). Combinatorial identities. A standardized set of tables listing 500 binomial coefficient summations. Morgantown, W.Va.

44. *Goulden I.P. and Jackson D.M.* (1983). Combinatorial enumeration. John Wiley, New York.
45. *Han H.S.W. and Reidys C.M.* (2008). Pseudoknot RNA structures with arc-length ≥ 4. J. Comput. Biol. **15**, no. 9, 1195–1208.
46. *Hardy G.H.* (1949). Divergent series. Clarendon Press, Oxford.
47. *Henrici P.* (1991). Applied and computational complex analysis. John Wiley, New York.
48. *Howe R.* (1974). The Fourier transform and germs of characters (case of Gl_n over p-adic field). Math. Ann. **208**, 305–322.
49. *Huang I-Ch.* (1997). Applications of residues to combinatorial identities. Proc. Amer. Math. Soc. **125**: 4, 1011–1017.
50. *Huang I-Ch.* (1998). Reversion of power series by residues. Comm. Algebra, **26**, 803–812.
51. *Huang I-Ch.* (2002). Inverse relations and Schauder bases. J. Combin. Theory Series A **97**, 203–224.
52. *Joni S.A.* (1978). Lagrange inversion in higher dimensions and umbral operators. Lin. and Mult. Algebra **6**, 111–121.
53. *Krattenthaler Ch.* (1984). A new q-Lagrange formula and some applications. Proc. Amer. Math. Soc. **90**, 338–344.
54. *Krattenthaler Ch.* (1988). Operator methods and Lagrange inversion: a unified approach to Lagrange formulas. Trans. Amer. Math. Soc. **305**, 431–465.
55. *Krattenthaler Ch.* (1996). A new matrix inverse. Proc. Amer. Math. Soc. **124**, 47–59.
56. *Krattenthaler Ch. and Schlosser M.* (1999). A new multidimensional matrix inverse with applications to multiple q-series. Discrete Math. **204**, 249–279.
57. *Krivokolesko V.P. and Tsikh A.K.* (2005). Integral representations in linearly convex polyhedra. Sib. Math. Journal **46**: 3, 579–593 (in Russian).
58. *Krivokolesko V.P.* (2008). About an integral representation, 59 pages (to appear).
59. *Kurosh A.G.* (1973). Lecture of general algebra. Nauka, M. (in Russian).
60. *Lagarias J.E.* (1997). The $3x+1$ Problem and its Generalizations, In: Borwein J. et al. (Eds.), Organic mathematics. Proc. workshop Simon Fraser Univ., Barnaby, Canada, Dec. 12-14, 1995; AMS, Providence, RI, 305–334.
61. *Leinartas E.K.* (1989). The Hadamard multidimensional composition and sums with linear constraints on summation indices. Sib. Mat. J. **30**: 4, 102–107 (in Russian).
62. *Leinartas E.K.* (2006). Integral methods in multiple theory of power series and difference equations. PhD thesis, Krasnoyarsk State Univ., Krasnoyarsk, 156 pages (in Russian).
63. *Leont'ev A.F.* (1980). The sequences of exponential polynomials. Nauka, M. (in Russian).
64. *Leont'ev V.K.* (2006). On the roots of random polynomials over a finite field. Math. Zametki, **80**: 2, 300–304 (in Russian).
65. *Liu, Y.* (1999). Enumerative theory of maps. Mathematics and its Applications, **468**. Kluwer Academic Publishers, Dordrecht; Science Press, Beijing.
66. *López B., Marco J.M. and Parcet J.* (2006) Taylor series and the Askey – Wilson operator and classical summation formulas. Proc. Amer. Math. Soc. **134**: 8, 2259–2270.
67. *Lushnikov A.A.* (2005). Exact kinetics of the sol-gel transition. Phys. Rev. E **71**, 0406129-1–0406129-10.
68. *Lushnikov A.A.* (2006). Gelation in coagulating systems. Phys. D **222**, 37–53.
69. *MacMahon P.A.* (1915–1916). Combinatory analysis. Vol. **I, II**. Cambridge Univ. Press.
70. *Mandelbrojt S.* (1973). Séries de Dirichlet. Principes et méthodes. Mir, Moscow (in Russian).
71. *Margenstern M. and Matiyasevich Y.* (1999). A binomial representation of the $3x+1$ problem, Acta Arith. **91**, 367–378.
72. *Marichev O.I.* (1983). Handbook of integral transforms of higher transcendental functions. Theory and algorithmic tables. Ellis Horwood Limited.
73. *Materov E.N. and Yuzhakov A.P.* (2000). The Bott formula for toric varieties and some combinatorial identities. Complex analysis and differential operators, Krasnoyarsk, 85–92.
74. *Materov E.N.* (2002). The Bott formula for toric varieties. Mosc. Math. J. **2**, no. 1, 161–182, 200.
75. *Merlini D., Sprugnoli R. and Verri M.C.* (2007). The method of coefficients. Amer. Math. Monthly **114**, 40–57.

76. *Mitrinović D.S. and Kečkić J.D.* (1984). Cauchy method of residues. Theory and applications. Vol. **I**, **II**. Kluwer Acad Press.
77. *Morrison K.E.* (2006). Integer Sequences and Matrices Over Finite Fields. J. Integer Sequences **9**, Article 06.2.1, 28 pages.
78. *Odlyzko A.M.* (1995). Asymptotic enumeration methods. Handbook of combinatorics, Vol. **1**, **2**, 1063–1229, Elsevier, Amsterdam.
79. *Paule P.* (1990). Computer Algebra Algorithmen für q-Reihen und kombinatorische Identitaten. RISC Linz, No 90-02.0, .25 pages.
80. *Paule P.* (1995). Greatest factorial factorization and symbolic summation. J. Symbolic Comput. **20**, 235–268.
81. *Petkovšek M., Wilf H.S. and Zeilberger D.* (1996). $A = B$. A K Peters, Wellesley, MA.
82. *Plamenevskii B.A.* (1986). Algebras of pseudo-differential operators. Nauka, M., 256 pages (in Russian).
83. *Pollard J.M.* (1971). The Fast Fourier Transform in a Finite Field. Math. Comp. **25**: 365–374.
84. *Pólya G.* (1937). Kombinatorische Anzahlbestimmungen für Gruppen, Graphen und chemische Verbindungen. Acta Math. **68**, 145–254.
85. *Postnikov M.M.* (1963). Foundations of Galois theory. Fizmatlit, Moscow (in Russian).
86. *Prudnikov A.P., Brychkov Yu.A. and Marichev O.M.* (1988). Integrals and Rings. Special functions. Vol. **1**. John Wiley, Berlin.
87. *Rademacher H.* (1973). Topics in Analytic Number Theory. Springer Verlag, New York.
88. *Ramakrishan D. and Valenza R.J.* (1999). Fourier analysis on number fields. Graduate Texts in Mathematics, **186**. Springer-Verlag, New York.
89. *Riordan J.* (1968). Combinatorial identities. John Wiley.
90. *Roman S.* (1984). The umbral calculus. Pure and Applied Mathematics, **111**. Acad. Press, New York.
91. *Rota G.-C.* (1964). On the foundations of combinatorial theory. I. Theory of Möbius functions. Z. Wahrsch. Verw. Gebiete **2**, 340–368.
92. *Sadykov T.M.* (2009). Hypergeometric functions of many complex variables. PhD thesis, Siberian Federal Univ., Krasnoyarsk, 261 pages (in Russian).
93. *Samko S.G., Kilbas A.A. and Marichev O.I.* (1993). Fractional integrals and derivatives. Theory and applications. Gordon and Breach, Yverdon.
94. *Schwatt I.J.* (1962). An introduction to the operations with series. Second edition, Chelsea Publishing Co., New York.
95. *Shabat B.V.* (1969). An introduction to the complex analysis, Nauka, M. (in Russian).
96. *Shapiro L.W., Getu S., Woan W.J. and Woodson L.C.* (1991). The Riordan group. Discrete Appl. Math. **34**, 229–239.
97. *Sloane N.J.A. and Plouffe S.* (1995). The encyclopedia of integer sequences. Acad. Press, San Diego.
98. *Sprugnoli R.* (2006). An introduction to mathematical methods in combinatorics. Dipartimento di Sistemi e Informatica Viale Morgagni, 65 – Firenze (Italy), 100 pages.
99. *Stanley R.P.* (1997, 1999). Enumerative combinatorics: Vol. **I**, **II**. Cambridge Univ. Press, Cambridge.
100. *Stepanenko V.A.* (2003). On the solution of the system of n algebraic equations with n variables with the help of hypergeometric functions. Vestnik Krasnoyarsk State Univ. **2**, Krasnoyarsk, 35–48 (in Russian).
101. *Stepanenko V.A.* (2005). Systems of algebraic equations, hypergeometric functions and integrals of several rational differentials. PhD thesis, Krasnoyarsk State Univ., Krasnoyarsk, 81 pages (in Russian).
102. *Stepanenko V.A.* (2008). Further chapters of mathematical analysis. Siberian Federal Univ., Krasnoyarsk, 176 pages (in press).
103. *Sun Y.* (2004). The statistic "number of udu's" in Dyck paths, Discrete Math. **287**, 177–186.
104. *Sun Z.W. and Davis M.* (2007). Combinatorial congruences modulo prime powers. Trans. Amer. Math. Soc. **359**: 11, 5525–5553.

105. *Tefera A.* (2002). MultInt, a MAPLE Package for Multiple Integration by the WZ Method. J. Symbolic Comput. **34**, 329–353.
106. *Titchmarsh E.C.* (1951). The theory of the Riemann Zeta-function, Oxford.
107. *Tsikh A.K.* (1992). Multidimensional residues and their applications. Translations of Mathematical Monographs, **103**, AMS, Providence, RI.
108. *Ufnarovsky V.A.* (1995). Combinatorial and asymptotic methods in algebra. Algebra, VI, 1–196, Encyclopaedia Math. Sci., 57, Springer, Berlin.
109. *Wang W. and Wang T.* (2009). Identities on Bell polynomials and Sheffer sequences. Discrete Math., **309**, no. 6, 1637–1648.
110. *Whiteman A.L.* (1953). Finite Fourier Series and equations in finite fields. Trans. Amer. Math. Soc. **74**, 78–98.
111. *Wilf H.S.* (1989). The "Snake-Oil" method for proving combinatorial identities. Surveys in combinatorics, London Math. Soc., Lecture Note Ser. **141**, Cambridge Univ. Press, Cambridge, 208–217.
112. *Wirsching G.J.* (1998). The dynamic system generated by the $3n+1$ function. Lecture Notes in Math. **1681**, Springer-Verlag, Berlin.
113. *Woodcock C.F.* (1996). Special p-adic analytic functions and Fourier transforms. J. Theory Numbers **60**, 393–408.
114. *Xin G.* (2005). A residue theorem for Malcev – Neumann series, Adv. in Appl. Math. **35**, 271–293.
115. *Xin G.* (2004). The ring of Malcev – Neumann series and the residue theorem, PhD thesis, Brandeis University, Waltham, MA, USA.
116. *Zeilberger D.* (1991). The method of creative telescoping. J. Symbolic Comput. **11**, 195–204.

Chapter 2
Partitions With Distinct Evens

George E. Andrews

Abstract Partitions with no repeated even parts (DE-partitions) are considered. A DE-rank for DE-partitions is defined to be the integer part of half the largest part minus the number of even parts. $\Delta(n)$ denotes the excess of the number of DE-partitions with even DE-rank over those with odd DE-rank. Surprisingly $\Delta(n)$ is (1) always non-negative, (2) almost always zero, and (3) assumes every positive integer value infinitely often. The main results follow from the work of Corson, Favero, Liesinger and Zubairy. Companion theorems for DE-partitions counted by exceptional parts conclude the paper.

2.1 Introduction

In [4] Ramanujan's series [16, p. 14]

$$R(q) = \sum_{n=0}^{\infty} \frac{q^{n(n+1)/2}}{(-q;q)_n} = \sum_{n=0}^{\infty} S(n)q^n \tag{2.1}$$

was examined. Here

$$(A;q)_n = (1-A)(1-Aq)\cdots(1-Aq^{n-1}). \tag{2.2}$$

It was shown [4] that $S(n)$ is almost always equal to zero and also assumes every integral value infinitely often. Combinatorially $S(n)$ is the excess of the number of partitions of n into distinct parts with even rank over those with odd rank. The *rank*

George E. Andrews
Department of Mathematics, The Pennsylvania State University, University Park, PA 16802, e-mail: andrews@math.psu.edu
Partially supported by National Science Foundation Grant DMS 0457003

I.S. Kotsireas, E.V. Zima (eds.), *Advances in Combinatorial Mathematics*,
DOI 10.1007/978-3-642-03562-3_2, © Springer-Verlag Berlin Heidelberg 2009

of a partition is the largest part minus the number of parts [7], [5]. A similar theorem was proven [4, Sec. 5] for partitions into odd parts without gaps.

The results for $S(n)$ rely crucially on the identity [4, p. 392]

$$R(q) = \sum_{\substack{n \geqq 0 \\ |j|<n}} (-1)^{n+j} q^{n(3n+1)/2-j^2}(1-q^{2n+1}). \tag{2.3}$$

It was noted at the end of [4] that there are numerous series similar in form to the right-hand expression in (2.3).

Indeed, results of this nature were given for Ramanujan's fifth order mock theta functions [2] (c.f. [17]), and such identities formed the basis for pathbreaking work by Zwegers [18] and Bringmann, Ono and Rhoades [6].

The object of this paper is to reveal a similar phenomenon connected to DE-partitions, i.e. partitions with no repeated even parts. Now DE-partitions have been examined previously. R. Honsberger [13] proved the following Euler-type theorem.

Theorem 2.1. *Let $P_{DE}(n)$ denote the number of partitions of n with no repeated even parts. Let $P_{<4}(n)$ denote the number of partitions of n in which no part appears more than thrice. Let $P_{\nmid 4}(n)$ denote the number of partitions of n into parts not divisible by 4. Then*

$$P_{DE}(n) = P_{<4}(n) = P_{\nmid 4}(n)$$

for each $n \geqq 0$.

Honsberger's proof is immediate from the following identification of the related generating functions

$$\begin{aligned}
\sum_{n \geqq 0} P_{DE}(n) q^n &= \frac{(-q^2;q^2)_\infty}{(q;q^2)_\infty} = \frac{(q^4;q^4)_\infty}{(q;q^2)_\infty (q^2;q^2)_\infty} \\
&= \frac{(q^4;q^4)_\infty}{(q;q)_\infty} = \sum_{n \geqq 0} P_{\nmid 4}(n) q^n \\
&= \prod_{n=1}^{\infty} (1+q^n+q^{2n}+q^{3n}) = \sum_{n \geqq 0} P_{<4}(n) q^n.
\end{aligned}$$

The fact that $P_{<4}(n) = P_{\nmid 4}(n)$ is due to J. W. L. Glaisher [10], and the asymptotics of these partition functions has been completely examined by P. Hagis [11].

From here on, our focus will be on the DE-rank of DE-partitions which is defined to be the integer part of half the largest part minus the number of even parts.

We let $\delta(m,n)$ denote the number of DE-partitions of n with DE-rank m.

Theorem 2.2.

$$\sum_{m,n \geqq 0} \delta(m,n) z^m q^n = 1 + \sum_{j \geqq 0} \frac{(-z^{-1}q^2;q^2)_j z^j q^{2j+1}(1+q)}{(q;q^2)_{j+1}}. \tag{2.4}$$

Next we write

$$\Delta(n) = \sum_{m \geq 0} (-1)^m \delta(m,n). \tag{2.5}$$

Theorem 2.3.

$$\sum_{n\geq 0} \Delta(n)(-q)^n = \sum_{n\geq 0} \frac{(-1)^n q^{n(n+1)/2}(q;q)_n}{(-q)_n}. \tag{2.6}$$

Fortunately, the expression on the right-hand side of (2.6) is, in fact, $W_1(-q)$, a function studied extensively by Corson et al. in [7]. In particular, their Theorem 2.3 combined with our Theorem 2.3 yields

Theorem 2.4.

$$\sum_{n\geq 0} \Delta(n)q^n = \sum_{n=0}^{\infty} \left(q^{\binom{2n+1}{2}} + q^{\binom{2n+2}{2}}\right) \sum_{j=-n}^{n} q^{-j^2}. \tag{2.7}$$

Theorem 3.2 of Corson et al. [7], may be restated here as:

Theorem 2.5. *$\Delta(n)$ is the number of inequivalent elements of the ring of integers of $Q(\sqrt{2})$ with norm $8n+1$.*

It immediately follows that

Corollary 2.1. *$\Delta(n)$ is always non-negative.*

Finally, Corson et al. [7] in the Remark just before their Corollary 5.3 make an assertion equivalent to

Corollary 2.2. *$\Delta(n)$ is almost always equal to zero.*

The Corollary 5.3 of Corson et al. [7] is equivalent to

Corollary 2.3. *$\Delta(n)$ is equal to any given positive integer infinitely often.*

The above results in some sense relate only half of the Corson et al. [7] paper to DE-partitions. In order to consider their companion function $W_2(q)$, we need a new definition related to DE-partitions. We shall say that a part of a DE-partition is *exceptional* if it is either even or one of the smallest parts or both. For example, $5+4+2+1+1$ is a DE-partition with four exceptional parts.

We let $\varepsilon(m,n)$ denote the number of DE-partitions of n with m exceptional parts, and we write

$$E(n) = \sum_{m\geq 0} (-1)^{m-1} \varepsilon(m,n). \tag{2.8}$$

Our main result for $E(n)$ requires the $W_2(q)$ of Corson et al. [7]:

$$W_2(q) = \sum_{n=1}^{\infty} \frac{(-1;q^2)_n(-q)^n}{(q;q^2)_n}. \tag{2.9}$$

Theorem 2.6.

$$\sum_{n=1}^{\infty} E(n)q^n = W_2(-q) - \sum_{n=1}^{\infty} q^{\binom{n+1}{2}}. \tag{2.10}$$

This assertion allows us to utilize Theorem 3.3 of Corson et al. [7] to establish immediately that

Theorem 2.7.

$$\sum_{n=1}^{\infty} E(n)q^n = \sum_{n\geq 1} \left(q^{\binom{2n}{2}} + q^{\binom{2n+1}{2}}\right) \sum_{\substack{j=-n \\ j\neq 0}}^{n-1} q^{-j^2-j}.$$

The three results following Theorem 2.4 now have perfect analogs as consequences of Theorem 2.7. These follow for Theorem 3.3 of [7], the Remark preceeding Corollary 5.3 and the proof of Corollary 5.3.

Theorem 2.8. *$E(n)$ is the number of inequivalent elements of the ring of integers of $Q(\sqrt{2})$ with norm $8n-1$ or one less if n is a triangular number.*

Corollary 2.4. *$E(n)$ is always non-negative.*

Corollary 2.5. *$E(n)$ is almost always equal to 0.*

The analog of Corollary 2.3 is quite plausible but it does not follow directly because of the second term in (2.10).

The remainder of the paper will be devoted to proofs of Theorems 2.2, 2.3 and 2.6. All the other results are, as noted, direct consequences of these three results and results in Corson et al. [7].

I thank Dean Hickerson for an extensive set of comments on this paper. In particular he has noted that $\Delta(n)$ is also the number of divisors of $8n+1$ which are congruent to ± 1 modulo 8 minus the number which are congruent to 3 or 5 modulo 8. Consequently, $\Delta(n)$ is the coefficient of $8n+1$ in

$$\sum_{\substack{n=1 \\ n \text{ odd}}}^{\infty} \frac{\left(\frac{2}{n}\right) q^n}{1-q^n},$$

where $\left(\frac{2}{n}\right)$ is the Legendre symbol.

Finally I note that A. Patkowski [15] has recently found two related theorems for DE-partitions. His theorems provide other lacunary series arising from DE-partition statistics other than the rank.

2.2 Proof of Theorem 2.2

For those DE-partitions with largest part $2j+1$, the DE-rank generating function is

$$\frac{(1+z^{-1}q^2)(1+z^{-1}q^4)\cdots(1+z^{-1}q^{2j})z^jq^{2j+1}}{(1-q)(1-q^3)\cdots(1-q^{2j+1})}.$$

For those DE-partitions with largest part $2j+2$, the DE-rank generating function is

$$\frac{(1+z^{-1}q^2)(1+z^{-1}q^4)\cdots(1+z^{-1}q^{2j})z^jq^{2j+2}z^{-1}}{(1-q)(1-q^3)\cdots(1-q^{2j+1})}.$$

We take the empty partition to have DE-rank 0, and so adding together the empty case, the odd case and the even case we find

$$\sum_{m,n\geqq 0}\delta(m,n)z^mq^n = 1+\sum_{j=0}^{\infty}\frac{(-z^{-1}q^2;q^2)_jz^j(q^{2j+1}+q^{2j+2})}{(q;q^2)_{j+1}},$$

which is equivalent to Theorem 2.2. □

2.3 Proof of Theorem 2.3

By Theorem 2.2 with z replaced by -1 and q replaced by $-q$, we see that

$$\begin{aligned}
\sum_{n\geqq 0}\Delta(n)(-q)^n &= \sum_{m,n\geqq 0}\delta(m,n)(-1)^{m+n}q^n \\
&= 1+\sum_{j\geqq 0}\frac{(q^2;q^2)_j(-1)^{j-1}q^{2j+1}(1-q)}{(-q;q^2)_{j+1}} \\
&= 1-\frac{q(1-q)}{1+q}\sum_{j\geqq 0}\frac{(q^2;q^2)_j(q^2;q^2)_j(-q^2)^j}{(q^2;q^2)_j(-q^3;q^2)_j} \\
&= 1+\sum_{j=0}^{\infty}\frac{(q;q^2)_{j+1}(-q)^{j+1}}{(-q^2;q^2)_{j+1}}
\end{aligned}$$

(by [9, eq. (III.2), p. 241 with $q\to q^2$, then $a=b=q^2$, $z=-q^2$, $c=-q^3$])

$$\begin{aligned}
&= \sum_{j=0}^{\infty}\frac{(q;q^2)_j(-q)^j}{(-q^2;q^2)_j} \\
&= \sum_{j\geqq 0}\frac{(q;q^2)_j(q^2;q^2)_j}{(-q^2;q^2)_j(-q;q^2)_j}(q^3)^jq^{2j^2-2j}(1+q^{4j+2})
\end{aligned}$$

(by [3, eq. (9.1.1), p. 223, $q\to q^2$, then $\alpha=q$, $\beta=-q^2$, $\tau=-q$])

$$= \sum_{j \geqq 0} \frac{(q;q)_{2j}}{(-q;q)_{2j+1}} q^{2j^2+j}(1+q^{2j+1} - q^{2j+1}(1-q^{2j+1}))$$

$$= \sum_{j=0}^{\infty} \frac{(q;q)_{2j} q^{\binom{2j+1}{2}}}{(-q;q)_{2j}} - \sum_{j=0}^{\infty} \frac{(q;q)_{2j+1} q^{\binom{2j+2}{2}}}{(-q;q)_{2j+1}}$$

$$= \sum_{n=0}^{\infty} \frac{(-1)^n q^{n(n+1)/2}(q;q)_n}{(-q;q)_n}. \qquad \square$$

2.4 Proof of Theorem 2.6

We require two results from the literature:

$$\sum_{n=0}^{\infty} q^{\binom{n+1}{2}} = \frac{(q^2;q^2)_\infty}{(q;q^2)_\infty} \qquad \text{[9, p. 6, eq. (7.321)]} \tag{2.11}$$

and

$$_2\phi_1\begin{pmatrix} a,b;q,t \\ c \end{pmatrix} = \frac{(\frac{abt}{c};q)_\infty}{(t;q)_\infty} {}_2\phi_1\begin{pmatrix} \frac{c}{a},\frac{c}{b};q,t \\ c \end{pmatrix} \qquad \text{[10, p. 10, weq. (1.4.6)]} \tag{2.12}$$

where

$$_2\phi_1\begin{pmatrix} a,b;q,t \\ c \end{pmatrix} = \sum_{n=0}^{\infty} \frac{(a;q)_n(b;q)_n t^n}{(q;q)_n(c;q)_n}. \tag{2.13}$$

Thus starting from (2.9)

$$W_2(-q)+1 = \sum_{n=0}^{\infty} \frac{(-1;q^2)_n q^n}{(-q;q^2)_n} = {}_2\phi_1\begin{pmatrix} -1,q^2;q^2,q \\ -q \end{pmatrix}$$

$$= \frac{(q^2;q^2)_\infty}{(q;q^2)_\infty} \sum_{n \geqq 0} \frac{(-q^{-1};q^2)_n(q;q^2)_n q^{2n}}{(q^2;q^2)_n(-q;q^2)_n} \qquad \text{by (2.12)}$$

Consequently

$$W_2(-q)+1-\frac{(q^2;q^2)_\infty}{(q;q^2)_\infty} = \sum_{n \geqq 1} \frac{(1+q^{-1})}{(1+q^{2n-1})} q^{2n} \frac{(q^{2n+2};q^2)_\infty}{(q^{2n+1};q^2)_\infty} \tag{2.14}$$

$$= \sum_{n \geqq 1} \frac{(1+q^{2n-1}+q^{-1}(1-q^{2n}))}{(1+q^{2n-1})} q^{2n} \frac{(q^{2n+2};q^2)_\infty}{(q^{2n+1};q^2)_\infty}$$

$$= \sum_{n \geqq 1} \frac{q^{2n}(q^{2n+2};q^2)_\infty}{(q^{2n+1};q^2)_\infty} + \sum_{n \geqq 1} \frac{q^{2n-1}(q^{2n};q^2)_\infty}{(1+q^{2n-1})(q^{2n+1};q^2)_\infty}.$$

Now the first sum above counts DE-partitions with smallest part even and a weight of $+1$ if there are an odd number of exceptional parts and -1 if there are an even number. The second sum counts DE-partitions with smallest part odd and a weight of $+1$ if there are an odd number of exceptional parts and -1 if there are an even number. Thus the right-hand side of (2.14) is the generating function for $E(n)$. Invoking (2.11), we see that

$$\sum_{n=1}^{\infty} E(n)q^n = W_2(-q) - \sum_{n=1}^{\infty} q^{\binom{n+1}{2}}. \qquad \square$$

2.5 Conclusion

There are a number of natural questions that arise from this study. First, combinatorial proofs of Theorems 2.4 and 2.7 might be possible and are much to be desired.

In addition, the ordinary rank of Dyson has led both to explications of the Ramanujan congruence for $p(n)$ (cf. [5] and [8]) and to surprising and appealing combinatorial theorems (cf. [9, eqs. (2.3.91) and (2.4.6)]. These aspects of the DE-rank and of exceptional parts of DE-partitions are completely unexplored.

References

1. G. E. Andrews, Multiple series Rogers-Ramanujan identities, *Pac. J. Math.*, **114** (1984), 267-283.
2. G. E. Andrews, The fifth and seventh order mock theta functions, *Trans. Amer. Math. Soc.*, **293** (1986), 113-134.
3. G. E. Andrews and B. Berndt, *Ramanujan's Lost Notebook, Part I*, Springer, New York, 2005.
4. G. E. Andrews, F. J. Dyson and D. Hickerson, Partitions and indefinite quadratic forms, *Inven. Math.*, **91** (1988), 391-407.
5. A. O. L. Atkin and H. P. F. Swinnerton-Dyer, Some properties of partitions, *Proc. London Math. Soc.* (3), **4** (1954), 84-106.
6. K. Bringmann, K. Ono and R. Rhoades, *J. Amer. Math. Soc.* (to appear).
7. D. Corson, D. Favero, K. Liesinger, and S. Zubairy, Characters and q-series in $\mathbb{Q}(\sqrt{2})$, *J. Number Th.*, **107** (2004), 392-405.
8. F. J. Dyson, Some guesses in the theory of partitions, *Eureka*, Feb. 1944, 10-15.
9. N. J. Fine, *Basic Hypereometric Seris and Applications*, Amer. Math. Soc., Providence, RI, 1988.
10. G. Gasper and M. Rahman, Basic Hypergeometric Series, *Encycl. Math. Applics.* Vol. 35, Cambridge University Press, Cambridge, 1990.
11. J. W. L. Glaisher, A theorem in partitions, *Messenger of Math.*, **12** (1883), 158-170.
12. P. Hagis, Partitions with a restriction on the multiplicity of summands, *Trans. Amer. Math. Soc.*, **155** (1971), 375-384.
13. R. Honsberger, Mathematical Gems III, *Math, Assoc. of Amer.*, Washington D.C., 1985.
14. W. J. LeVeque, *Topics in Number Theory*, Vol. I, Addison-Wesley, Reading, MA, 1956.
15. W. J. LeVeque, *Topics in Number Theory*, Vol. II, Addison-Wesley, Reading, MA, 1956.
16. A. Patkowski, On some partitions where even parts do not repeat, (to appear).
17. S. Ramanujan, *The Lost Notebook and Other Unpublished Papers*, Narosa, New Delhi, 1988.
18. S. Zwegers, Mock Theta Functions, Ph.D. thesis, Unversiteit Utrecht, 2002.

Chapter 3
A factorization theorem for classical group characters, with applications to plane partitions and rhombus tilings

M. Ciucu and C. Krattenthaler

Dedicated to Georgy P. Egorychev

Abstract We prove that a Schur function of rectangular shape (M^n) whose variables are specialized to $x_1, x_1^{-1}, \dots, x_n, x_n^{-1}$ factorizes into a product of two odd orthogonal characters of rectangular shape, one of which is evaluated at $-x_1, \dots, -x_n$, if M is even, while it factorizes into a product of a symplectic character and an even orthogonal character, both of rectangular shape, if M is odd. It is furthermore shown that the first factorization implies a factorization theorem for rhombus tilings of a hexagon, which has an equivalent formulation in terms of plane partitions. A similar factorization theorem is proven for the sum of two Schur functions of respective rectangular shapes (M^n) and (M^{n-1}).

3.1 Introduction

The purpose of this note is to prove curious factorization properties for Schur functions of rectangular shape, which seem to have escaped the attention of previous authors. (We refer the reader to Section 3.2 for all definitions.) More precisely, we show that a Schur function of rectangular shape (M^n) which is evaluated at $x_1, x_2, \dots, x_n$ and their reciprocals $x_1^{-1}, x_2^{-1}, \dots, x_n^{-1}$ factorizes into two factors, and

M. Ciucu
Department of Mathematics, Indiana University, Bloomington, IN 47405-5701, USA
C. Krattenthaler
Fakultät für Mathematik, Universität Wien, Nordbergstraße 15, A-1090 Vienna, Austria.
The research of the first author was partially supported by NSF grant DMS-0500616. The research of the second author was partially supported by the Austrian Science Foundation FWF, grants Z130-N13 and S9607-N13, the latter in the framework of the National Research Network "Analytic Combinatorics and Probabilistic Number Theory." This work was done during the authors' stay at the Erwin Schrödinger Institute for Physics and Mathematics, Vienna, during the programme "Combinatorics and Statistical Physics" in Spring 2008.

I.S. Kotsireas, E.V. Zima (eds.), *Advances in Combinatorial Mathematics*,
DOI 10.1007/978-3-642-03562-3_3, © Springer-Verlag Berlin Heidelberg 2009

the same is true for the sum of two Schur functions of respective shapes (M^n) and (M^{n-1}).

We begin by describing explicitly the case of one Schur function. If M is even, then both factors are odd orthogonal characters of rectangular shape, one of them evaluated at the variables $x_1, x_2, \ldots, x_n$, the other is evaluated at $-x_1, -x_2, \ldots, -x_n$. If M is odd, then one factor is a symplectic character of rectangular shape, while the other is an even orthogonal character of rectangular shape, both being evaluated at $x_1, x_2, \ldots, x_n$. The case of even M of this factorization property is presented in the following theorem.

Theorem 3.1. *For any non-negative integers m and n, we have*

$$s_{((2m)^n)}(x_1, x_1^{-1}, x_2, x_2^{-1}, \ldots, x_n, x_n^{-1}) \\ = (-1)^{mn}\, so_{(m^n)}(x_1, x_2, \ldots, x_n)\, so_{(m^n)}(-x_1, -x_2, \ldots, -x_n). \quad (3.1)$$

If M is odd, then the factorization takes the following form.

Theorem 3.2. *For any non-negative integers m and n, we have*

$$s_{((2m+1)^n)}(x_1, x_1^{-1}, x_2, x_2^{-1}, \ldots, x_n, x_n^{-1}) \\ = sp_{(m^n)}(x_1, x_2, \ldots, x_n)\, o^{even}_{((m+1)^n)}(x_1, x_2, \ldots, x_n). \quad (3.2)$$

Since our identities involve classical group characters, one might ask whether there are representation-theoretic interpretations of these identities. At first sight, this seems to be a difficult question because of the somewhat "incoherent" right-hand sides of (3.1) and (3.2). However, there is a uniform way of writing the factorization identities of Theorems 3.1 and 3.2 that was pointed out to us by Soichi Okada. Namely, by comparing (3.7) and (3.11), and by using (3.8) and (3.12), we see that

$$(-1)^{\sum_{i=1}^N \lambda_i}\, so_\lambda(-x_1, -x_2, \ldots, -x_N) \prod_{i=1}^N (x_i^{1/2} + x_i^{-1/2}) = o^{even}_{\lambda+\frac{1}{2}}(x_1, x_2 \ldots, x_N), \quad (3.3)$$

where $\lambda + \frac{1}{2}$ is short for $(\lambda_1 + \frac{1}{2}, \lambda_2 + \frac{1}{2}, \ldots, \lambda_N + \frac{1}{2})$. Furthermore, by comparing (3.7) and (3.9), and by using (3.8) and (3.10), we see that

$$sp_\lambda(x_1, x_2, \ldots, x_N) \prod_{i=1}^N (x_i^{1/2} + x_i^{-1/2}) = so_{\lambda+\frac{1}{2}}(x_1, x_2 \ldots, x_N). \quad (3.4)$$

Theorem 3.1 and 3.2 may therefore be uniformly stated as

$$\left(\prod_{i=1}^n (x_i^{1/2} + x_i^{-1/2})\right) s_{(M^n)}(x_1, x_1^{-1}, x_2, x_2^{-1}, \ldots, x_n, x_n^{-1}) \\ = so_{((\frac{M}{2})^n)}(x_1, x_2, \ldots, x_n)\, o^{even}_{((\frac{M+1}{2})^n)}(x_1, x_2, \ldots, x_n). \quad (3.5)$$

Written in this way, this may, in the end, lead to a representation-theoretic interpretation of these identities, although we must confess that we are not able to offer such an interpretation.

On the other hand, we are able to offer a *combinatorial* interpretation for Theorem 3.1. As we show in Section 3.5, if we specialize $x_1 = x_2 = \cdots = x_n = 1$ in Theorem 3.1, then one obtains a factorization theorem for rhombus tilings of a hexagon, which has also a natural, equivalent formulation as a factorization theorem for plane partitions (see (3.26)). It is, in fact, this factorization theorem which we observed first, and which formed the starting point of this work. We suspect that a more general factorization theorem for rhombus tilings is lurking behind. If there is a natural combinatorial interpretation of Theorem 3.2 is less clear. We make an attempt, also in Section 3.5, but we consider it not entirely satisfactory.

The case of the sum of two Schur functions of rectangular shapes is treated in Section 3.6. Namely, we show that there are very similar factorization theorems for the sum of two Schur functions of respective rectangular shapes (M^n) and (M^{n-1}) (see Theorems 3.3 and 3.4). The existence of these was pointed out to us by Ron King.

The proofs of Theorems 3.1 and 3.2 are given in Section 3.4. These proofs are based on an auxiliary identity which is established in Section 3.3 (see Lemma 3.1). The proofs of Theorems 3.3 and 3.4 are based on two further auxiliary identities of similar type, which are also presented and proved in Section 3.3 (see Lemmas 3.2 and 3.3).

3.2 Classical group characters

In this section we recall the definitions of the classical group characters involved in the factorizations in Theorems 3.1 and 3.2. We also briefly touch upon their significance in representation theory.

Given a partition $\lambda = (\lambda_1, \lambda_2, \dots, \lambda_N)$ (i.e., a non-increasing sequence of non-negative integers) the *Schur function* $s_\lambda(x_1, x_2, \dots, x_N)$ is defined by (see [4, p. 403, (A.4)], [7, Prop. 1.4.4], or [8, Ch. I, (3.1)])

$$s_\lambda(x_1, x_2, \dots, x_N) = \frac{\det\limits_{1\le h,t\le N}\left(x_h^{\lambda_t+N-t}\right)}{\det\limits_{1\le h,t\le N}\left(x_h^{N-t}\right)}. \tag{3.6}$$

It is not difficult to see that the denominator in (3.6) cancels out, so that any Schur function $s_\lambda(x_1, x_2, \dots, x_N)$ is in fact a *polynomial* in $x_1, x_2, \dots, x_N$, and is thus well-defined for any choice of the variables $x_1, x_2, \dots, x_N$. It is well-known (cf. [4, §24.2]) that $s_\lambda(x_1, x_2, \dots, x_N)$ is an irreducible character of $SL_N(\mathbb{C})$ (respectively $GL_N(\mathbb{C})$).

Given a non-increasing sequence $\lambda = (\lambda_1, \lambda_2, \dots, \lambda_n)$ of integers or half-integers (the latter being, by definition, positive odd integers divided by 2), the *odd orthogonal character* $so_\lambda(x_1, x_2, \dots, x_N)$ is defined by (see [4, (24.28)])

$$so_{\lambda}(x_1,x_2,\dots,x_N) = \frac{\det\limits_{1\le h,t\le N}\left(x_h^{\lambda_t+N-t+\frac{1}{2}} - x_h^{-(\lambda_t+N-t+\frac{1}{2})}\right)}{\det\limits_{1\le h,t\le N}\left(x_h^{N-t+\frac{1}{2}} - x_h^{-(N-t+\frac{1}{2})}\right)}. \tag{3.7}$$

Again, it is not difficult to see that the denominator in (3.7) cancels out, so that any odd orthogonal character $so_{\lambda}(x_1,x_2,\dots,x_N)$ is in fact a *Laurent polynomial* in $x_1,x_2,\dots,x_N$ (i.e., a polynomial in $x_1,x_1^{-1},x_2,x_2^{-1},\dots,x_N,x_N^{-1}$), and is thus well-defined for any choice of the variables $x_1,x_2,\dots,x_N$ such that all of them are non-zero. By the *Weyl denominator formula for type B* (cf. [4, Lemma 24.3, Ex. A.62]),

$$\det_{1\le h,t\le n}\left(x_h^{n-t+\frac{1}{2}} - x_h^{-(n-t-\frac{1}{2})}\right)$$
$$= (x_1x_2\cdots x_n)^{-n+\frac{1}{2}} \prod_{1\le h<t\le n}(x_h-x_t)(x_hx_t-1)\prod_{h=1}^{n}(x_h-1), \tag{3.8}$$

the denominator in (3.7) can be actually evaluated in product form. It is well-known (cf. [4, §24.2]) that $so_{\lambda}(x_1,x_2,\dots,x_N)$ is an irreducible character of $SO_{2N+1}(\mathbb{C})$.

Given a partition $\lambda=(\lambda_1,\lambda_2,\dots,\lambda_n)$, the *symplectic character* $sp_{\lambda}(x_1,x_2,\dots,x_N)$ is defined by (see [4, (24.18)])

$$sp_{\lambda}(x_1,x_2,\dots,x_N) = \frac{\det\limits_{1\le h,t\le N}\left(x_h^{\lambda_t+N-t+1} - x_h^{-(\lambda_t+N-t+1)}\right)}{\det\limits_{1\le h,t\le N}\left(x_h^{N-t+1} - x_h^{-(N-t+1)}\right)}. \tag{3.9}$$

Similarly to odd orthogonal characters, $sp_{\lambda}(x_1,x_2,\dots,x_N)$ is a *Laurent polynomial* in $x_1,x_2,\dots,x_N$, and is thus well-defined for any choice of the variables $x_1,x_2,\dots,x_N$ such that all of them are non-zero. By the *Weyl denominator formula for type C* (cf. [4, Lemma 24.3, Ex. A.52]),

$$\det_{1\le h,t\le n}\left(x_h^{n-t+1} - x_h^{n-t+1}\right) = (x_1\cdots x_n)^{-n} \prod_{1\le h<t\le n}(x_h-x_t)(x_hx_t-1)\prod_{h=1}^{n}(x_h^2-1), \tag{3.10}$$

the denominator in (3.9) can be actually evaluated in product form. Furthermore, $sp_{\lambda}(x_1,x_2,\dots,x_N)$ is an irreducible character of $Sp_{2N}(\mathbb{C})$ (cf. [4, §24.2]).

Finally, given a non-increasing sequence $\lambda=(\lambda_1,\lambda_2,\dots,\lambda_N)$ of *positive* integers or half-integers, the *even orthogonal character* $o_{\lambda}^{even}(x_1,x_2,\dots,x_N)$ is given by (see [4, (24.40) plus the remarks on top of page 411])

$$o_{\lambda}^{even}(x_1,x_2,\dots,x_N) = 2\frac{\det\limits_{1\le h,t\le N}\left(x_h^{\lambda_t+N-t} + x_h^{-(\lambda_t+N-t)}\right)}{\det\limits_{1\le h,t\le N}\left(x_h^{N-t} + x_h^{-(N-t)}\right)}. \tag{3.11}$$

Here as well, $o_\lambda^{even}(x_1,x_2,\dots,x_N)$ is a *Laurent polynomial* in $x_1,x_2,\dots,x_N$, and is thus well-defined for any choice of the variables $x_1,x_2,\dots,x_N$ such that all of them are non-zero. the *Weyl denominator formula for type D* (cf. [4, Lemma 24.3, Ex. A.66]),

$$\det_{1\le h,t\le n}\left(x_h^{n-t}+x_h^{-(n-t)}\right)=2\cdot(x_1\cdots x_n)^{-n+1}\prod_{1\le h<t\le n}(x_h-x_t)(x_hx_t-1), \tag{3.12}$$

the denominator in (3.11) can be again evaluated in product form. It is well-known (cf. [4, §24.2]) that $o_\lambda^{even}(x_1,x_2,\dots,x_N)$ is an irreducible character of $O_{2N}(\mathbb{C})$. When restricted to $SO_{2N}(\mathbb{C})$, it splits into two different irreducible characters of $SO_{2N}(\mathbb{C})$. (The reader should observe that we assumed $\lambda_N>0$. If we had allowed $\lambda_N=0$, then we would have to divide the right-hand side of (3.11) by 2 in order to obtain an irreducible character of $O_{2N}(\mathbb{C})$ or $SO_{2N}(\mathbb{C})$.)

3.3 Auxiliary identities

The proofs of Theorems 3.1 and 3.2, and those of Theorems 3.3 and 3.4 in Section 3.6, hinge upon certain multivariable identities which we prove in this section. Before we are able to state these identities, we have to introduce some notation.

Throughout, we use the standard notation $[n]:=\{1,2,\dots,n\}$. Let A and B be given subsets of the set of positive integers. Slightly abusing notation for resultants from [7], we define

$$R\left(A,B^{-1}\right):=\prod_{a\in A}\prod_{b\in B}(x_a-x_b^{-1}).$$

(In order to avoid any confusion on the part of the reader: the symbol B^{-1} in $R(A,B^{-1})$ has no meaning by itself, the exponent -1 is just there to indicate that the reciprocals of the variables indexed by B are used on the right-hand side of the definition.) Furthermore, we define

$$V(A):=\prod_{\substack{a,b\in A\\ a<b}}(x_a-x_b)\quad\text{and}\quad V\left(A^{-1}\right):=\prod_{\substack{a,b\in A\\ a<b}}(x_a^{-1}-x_b^{-1}).$$

In all these definitions, empty products have to be interpreted as 1.

Now we are in the position to state and prove the identity on which the proofs of Theorems 3.1 and 3.2 are based.

Lemma 3.1. *For all positive integers N, there holds the identity*

$$\sum_{\substack{A\subseteq[2N]\\ |A|=N}}V(A)V\left(A^{-1}\right)V(A^c)V\left((A^c)^{-1}\right)R\left(A,A^{-1}\right)R\left(A^c,(A^c)^{-1}\right)$$
$$=\sum_{A\subseteq[2N]}V(A)V\left(A^{-1}\right)V(A^c)V\left((A^c)^{-1}\right)R\left(A,(A^c)^{-1}\right)R\left(A^c,A^{-1}\right), \tag{3.13}$$

where A^c denotes the complement of A in $[2N]$.

Proof. We prove the assertion by induction on N. For $N = 1$, identity (3.13) reduces to

$$2(x_1 - x_1^{-1})(x_2 - x_2^{-1}) = 2(x_1 - x_2)(x_1^{-1} - x_2^{-1}) + 2(x_1 - x_2^{-1})(x_2 - x_1^{-1}),$$

which can be readily verified.

Now let us suppose that we have proved (3.13) with N replaced by $N-1$. Below, we shall show that this implies that the identity (3.13) is true if we specialize x_1 to x_2. Let us for the moment suppose that this is already done. Since, as is easy to see, both sides of (3.13) are symmetric in the variables $x_1, x_2, \ldots, x_{2N}$, as well as they remain invariant up to an overall multiplicative sign if we replace x_1 by x_1^{-1} (on the left-hand side of (3.13), the latter assertion already applies to each individual summand; on the right-hand side, one has to group the summands in pairs: the summands corresponding to a set A and the symmetric difference $A \triangle \{1\}$ have to be considered together), this means that identity (3.13) holds for *all* specializations of x_1 to one of $x_2, x_2^{-1}, x_3, x_3^{-1}, \ldots, x_{2N}, x_{2N}^{-1}$. These are $4N-2$ specializations. The previously observed fact that the replacement of x_1 by x_1^{-1} changes the sign of both sides of (3.13) implies that both sides vanish for $x_1 = \pm 1$. Thus, we have $4N$ specializations of x_1 for which the two sides of (3.13) agree. On the other hand, as Laurent polynomials in x_1, the degree of both sides of (3.13) is at most $2N-1$. (That is, the maximal exponent e of a power x_1^e is $e = 2N-1$, and the minimal exponent is $e = -(2N-1)$.) Hence, both sides of (3.13) must be identical.

It remains to prove that, under the induction hypothesis, Equation (3.13) holds for $x_1 = x_2$. Indeed, under this specialization, terms corresponding to sets A which contain *both* 1 and 2 and to those which contain *neither* 1 *nor* 2 vanish on both sides of (3.13), because of the appearance of the Vandermonde products $V(A)$ respectively $V(A^c)$. Therefore, if $x_1 = x_2$, identity (3.13) reduces to

$$\begin{aligned}
&2(x_1 - x_1^{-1})^2 \left(\prod_{j=3}^{2N} (x_1 - x_j)(x_1^{-1} - x_j^{-1})(x_1 - x_j^{-1})(x_j - x_1^{-1}) \right) \\
&\quad \times \sum_{\substack{A \subseteq \{3,4,\ldots,2N\} \\ |A| = N-1}} V(A) V(A^{-1}) \, V(A^c) V((A^c)^{-1}) \, R(A, A^{-1}) \, R(A^c, (A^c)^{-1}) \\
&\quad = 2(x_1 - x_1^{-1})^2 \left(\prod_{j=3}^{2N} (x_1 - x_j)(x_1^{-1} - x_j^{-1})(x_1 - x_j^{-1})(x_j - x_1^{-1}) \right) \\
&\quad \times \sum_{A \subseteq \{3,4,\ldots,2N\}} V(A) V(A^{-1}) \, V(A^c) V((A^c)^{-1}) \, R(A, (A^c)^{-1}) \, R(A^c, A^{-1}) .
\end{aligned}$$

After clearing the product which is common to both sides, we see that the remaining identity is equivalent to (3.13) with N replaced by $N-1$, the latter being true due to the induction hypothesis. This finishes the proof of the lemma.

On the other hand, the proofs of Theorems 3.3 and 3.4 are based on the following two lemmas. The reader should observe the subtle difference on the right-hand sides of (3.13) and (3.14) (the left-hand sides being identical).

Lemma 3.2. *For all positive integers N, there holds the identity*

$$\sum_{\substack{A\subseteq[2N]\\|A|=N}} V(A)V(A^{-1})\,V(A^c)V((A^c)^{-1})\,R(A,A^{-1})\,R(A^c,(A^c)^{-1})$$
$$=\sum_{A\subseteq[2N]} \mathbf{x}^{-A}\mathbf{x}^{A^c}V(A)V(A^{-1})\,V(A^c)V((A^c)^{-1})\,R(A,(A^c)^{-1})\,R(A^c,A^{-1}),\quad (3.14)$$

where A^c denotes the complement of A in $[2N]$, and where $\mathbf{x}^{-A}$ is short for $\prod_{a\in A}x_a^{-1}$ and $\mathbf{x}^{A^c}$ is short for $\prod_{a\in A^c}x_a$.

Proof. We proceed as in the proof of Lemma 3.1. That is, we perform an induction on N. For $N=1$, identity (3.14) reduces to

$$2(x_1-x_1^{-1})(x_2-x_2^{-1})=x_1x_2(x_1-x_2)(x_1^{-1}-x_2^{-1})+x_1x_2^{-1}(x_1-x_2^{-1})(x_2-x_1^{-1})$$
$$+x_1^{-1}x_2(x_2-x_1^{-1})(x_1-x_2^{-1})+x_1^{-1}x_2^{-1}(x_1^{-1}-x_2^{-1})(x_1-x_2),$$

which can be readily verified. Almost all the remaining steps are identical with those in the proof of Lemma 3.1, except that more care is needed to show that, as a Laurent polynomial in x_1, the degree of the right-hand side of (3.14) is at most $2N-1$. Indeed, by inspection, this degree is at most $2N$, and the coefficient of x_1^{2N} is equal to

$$\sum_{A\subseteq[2N]\backslash\{1\}} (-1)^{|A^c|-1}V(A)V(A^{-1})\,V(A^c)V((A^c)^{-1})\,R(A,(A^c)^{-1})\,R(A^c,A^{-1}),$$

where A^c now denotes the complement of A in $[2N]\backslash\{1\}$. (Note that only those subsets A of $[2N]$ contribute to the coefficient of x_1^{2N} on the right-hand side of (3.14) which do not contain 1, whence the term $x^{-A}x^{A^c}$ on the right-hand side of (3.14) got cancelled due to the contributions from the terms $R(A,(A^c)^{-1})$ and $V((A^c)^{-1})$, respectively.) However, in this sum, the terms indexed by A respectively A^c cancel each other, so that this sum does indeed vanish.

Lemma 3.3. *For all positive integers N, there holds the identity*

$$\sum_{\substack{A\subseteq[2N+1]\\|A|=N}} V(A)V(A^{-1})\,V(A^c)V((A^c)^{-1})\,R(A,A^{-1})\,R(A^c,(A^c)^{-1})=$$
$$\sum_{A\subseteq[2N+1]} (-1)^{N+|A|}\mathbf{x}^{-A}\mathbf{x}^{A^c}V(A)V(A^{-1})\,V(A^c)V((A^c)^{-1})\,R(A,(A^c)^{-1})\,R(A^c,A^{-1}),$$
$$(3.15)$$

where A^c denotes the complement of A in $[2N+1]$, while $\mathbf{x}^{-A}$ is short for $\prod_{a\in A}x_a^{-1}$ and $\mathbf{x}^{A^c}$ is short for $\prod_{a\in A^c}x_a$, as before.

Proof. We proceed again as in the proof of Lemma 3.1. Here, the induction basis, the case $N=0$ of (3.15), reads

$$(x_1 - x_1^{-1}) = x_1 - x_1^{-1}.$$

As Laurent polynomials in x_1, the degree of the left-hand and right-hand sides of (3.15) are at most $2N+1$. This means that we need $4N+3$ specializations, respectively "informations," which agree on both sides of (3.15), in order to show that both sides are equal.

In the same way as this is done in Lemma 3.1, one can show that, if one assumes the truth of (3.15) with N replaced by $N-1$, this implies that (3.15) is true if we specialize x_1 to x_2. Similarly to the proof of Lemma 3.1, it is easy to see that both sides of (3.15) are symmetric in the variables $x_1, x_2, \dots, x_{2N+1}$, and also that they remain invariant up to an overall multiplicative sign if we replace x_1 by x_1^{-1}. This means that identity (3.15) holds for *all* specializations of x_1 to one of $x_2, x_2^{-1}, x_3, x_3^{-1}, \dots, x_{2N+1}, x_{2N+1}^{-1}$. These are $4N$ specializations. We still need 3 additional specializations, respectively "informations," which agree on both sides of (3.15).

We get two more specializations by observing that both sides of (3.15) vanish for $x_1 = \pm 1$. For the left-hand side this is obvious because of the factors $R(A, A^{-1})$ respectively $R(A^c, (A^c)^{-1})$ appearing in the summand. On the right-hand side, the summands indexed by A and $A \triangle \{1\}$ cancel each other for $x_1 = \pm 1$.

Finally, the coefficients of x_1^{2N+1} on the left-hand and right-hand sides of (3.15) are respectively

$$\sum_{\substack{A \subseteq [2N+1] \setminus \{1\} \\ |A|=N}} (-1)^N V(A) V(A^{-1})\, V(A^c) V((A^c)^{-1})\, R(A, A^{-1})\, R(A^c, (A^c)^{-1})$$

and

$$\sum_{A \subseteq [2N+1] \setminus \{1\}} (-1)^N V(A) V(A^{-1})\, V(A^c) V((A^c)^{-1})\, R(A, (A^c)^{-1})\, R(A^c, A^{-1}),$$

where, in both cases, A^c now denotes the complement of A in $[2N+1] \setminus \{1\}$. The equality of these two sums was established in Lemma 3.1. This completes the proof of the lemma.

3.4 Proofs of theorems

This section is devoted to the proofs of Theorems 3.1 and 3.2. The idea is to substitute the determinantal definitions (3.6)–(3.11) of the characters into (3.1) respectively (3.2), expand the determinants in the numerators by using Laplace expansion respectively linearity of the determinant in the rows, evaluate the resulting simpler determinants by means of one of the Weyl denominator formulas, and reduce the resulting expressions. By collecting appropriate terms, it is then seen that both identities result from Lemma 3.1 in the preceding section.

Proof of Theorem 3.1. We start with the left-hand side of (3.1). By (3.6), we have

$$s_{((2m)^n)}(x_1,x_1^{-1},x_2,x_2^{-1},\dots,x_n,x_n^{-1}) = \frac{\det\limits_{1\le i,j\le 2n}\begin{pmatrix} x_h^{2m\chi(t\le n)+2n-t} & 1\le h\le n \\ x_{h-n}^{-(2m\chi(t\le n)+2n-t)} & n+1\le h\le 2n\end{pmatrix}}{V([n])V([n]^{-1})R([n],[n]^{-1})}, \tag{3.16}$$

where $\chi(\mathscr{S})=1$ if $\mathscr{S}$ is true and $\chi(\mathscr{S})=0$ otherwise. Here, we used the evaluation of the Vandermonde determinant (the Weyl denominator formula for type A; cf. [4, p. 400 and Lemma 24.3]) in the denominator. We now do a Laplace expansion of the determinant along the first n columns. Abbreviating the denominator on the right-hand side of (3.16) by $D_1(n)$, this leads to

$$\begin{aligned} &s_{((2m)^n)}(x_1,x_1^{-1},x_2,x_2^{-1},\dots,x_n,x_n^{-1}) \\ &= \frac{1}{D_1(n)} \sum_{\substack{A,B\subseteq[n]\\ |A|+|B|=n}} (-1)^{\left(\sum_{a\in A}a\right)+\left(\sum_{b\in B}b\right)+n|B|-\binom{n+1}{2}} \det M_1(A,B)\cdot\det M_2(A^c,B^c), \end{aligned} \tag{3.17}$$

where $M_1(A,B)$ is the $n\times n$ matrix

$$\begin{pmatrix} x_h^{2m+2n-t} & h\in A \\ x_h^{-(2m+2n-t)} & h\in B \end{pmatrix},$$

and $M_2(A^c,B^c)$ is the $n\times n$ matrix

$$\begin{pmatrix} x_h^{n-t} & h\in A^c \\ x_h^{-(n-t)} & h\in B^c \end{pmatrix},$$

with A^c denoting the complement of A in $[n]$, and an analogous meaning for B^c. All determinants in (3.17) are Vandermonde determinants, except for some trivial factors which can be taken out of the rows of $M_1(A,B)$ respectively $M_2(A,B)$, and can therefore be evaluated in product form. If we substitute the corresponding results, we obtain

$$\begin{aligned} &s_{((2m)^n)}(x_1,x_1^{-1},x_2,x_2^{-1},\dots,x_n,x_n^{-1}) \\ &= \frac{1}{D_1(n)} \sum_{\substack{A,B\subseteq[n]\\ |A|+|B|=n}} (-1)^{\left(\sum_{a\in A}a\right)+\left(\sum_{b\in B}b\right)+n|B|-\binom{n+1}{2}} \left(\prod_{a\in A} x_a\right)^{2m+n} \left(\prod_{b\in B} x_b^{-1}\right)^{2m+n} \\ &\qquad \cdot V(A)V(B^{-1})R(A,B^{-1})V(A^c)V\left((B^c)^{-1}\right)R\left(A^c,(B^c)^{-1}\right). \end{aligned} \tag{3.18}$$

Next we turn to the right-hand side of (3.1). By (3.7), we have

$$so_{(m^n)}(x_1,x_2,\dots,x_n) = \frac{\det_{1\le h,t\le n}\left(x_h^{m+n-t+\frac{1}{2}} - x_h^{-(m+n-t+\frac{1}{2})}\right)}{\det_{1\le h,t\le n}\left(x_h^{n-t+\frac{1}{2}} - x_h^{-(n-t+\frac{1}{2})}\right)}$$

$$= \frac{\det_{1\le h,t\le n}\left(x_h^{m+n-t+\frac{1}{2}} - x_h^{-(m+n-t+\frac{1}{2})}\right)}{V([n])\prod_{i=1}^n x_i^{-n+i}(x_i^{1/2}-x_i^{-1/2})\prod_{1\le i<j\le n}(x_i-x_j^{-1})}, \tag{3.19}$$

where we have used the Weyl denominator formula (3.8). We now use linearity of the determinant in the rows. Abbreviating the denominator on the right-hand side of (3.19) by $D_2(n)$, this leads to

$$so_{(m^n)}(x_1,x_2,\dots,x_n) = \frac{1}{D_2(n)}\sum_{A\subseteq[n]}(-1)^{\left(\sum_{a\in A}a\right)-\binom{|A|+1}{2}}\det M_3(A), \tag{3.20}$$

where $M_3(A)$ is the $n\times n$ matrix

$$\begin{pmatrix} x_h^{m+n-t+\frac{1}{2}} & h\in A \\ -x_h^{-(m+n-t+\frac{1}{2})} & h\in A^c \end{pmatrix},$$

with A^c denoting the complement of A in $[n]$, as before. All determinants in (3.20) are Vandermonde determinants, except for some trivial factors which can be taken out of the rows of $M_3(A)$, and can therefore be evaluated in product form. If we substitute the corresponding results, we obtain

$$so_{(m^n)}(x_1,x_2,\dots,x_n)$$

$$= \frac{1}{D_2(n)}\sum_{A\subseteq[n]}(-1)^{\left(\sum_{a\in A}a\right)-\binom{|A|}{2}+n}\left(\prod_{a\in A}x_a\right)^{m+\frac{1}{2}}\left(\prod_{a\in A^c}x_a^{-1}\right)^{m+\frac{1}{2}}$$

$$\cdot V(A)V\left((A^c)^{-1}\right)R\left(A,(A^c)^{-1}\right), \tag{3.21}$$

whence the product of the two characters on the right-hand side of (3.1) can be expanded in the form

$$
\begin{aligned}
&so_{(m^n)}(x_1,x_2,\dots,x_n)\,so_{(m^n)}(-x_1,-x_2,\dots,-x_n)\\
&=\frac{(-1)^{mn}}{D_1(n)}\sum_{A,B\subseteq[n]}(-1)^{\left(\sum_{a\in A}a\right)+\left(\sum_{b\in B}b\right)-\binom{|A|+1}{2}-\binom{|B|}{2}}\\
&\quad\cdot\Big(\prod_{a\in A}x_a\Big)^{m+\frac12}\Big(\prod_{a\in A^c}x_a^{-1}\Big)^{m+\frac12}\Big(\prod_{b\in B}x_b\Big)^{m+\frac12}\Big(\prod_{b\in B^c}x_b^{-1}\Big)^{m+\frac12}\\
&\quad\cdot V(A)V\left((A^c)^{-1}\right)R\big(A,(A^c)^{-1}\big)V(B)V\left((B^c)^{-1}\right)R\big(B,(B^c)^{-1}\big). \qquad (3.22)
\end{aligned}
$$

We now fix disjoint subsets A' and B' of $[n]$ and extract the coefficient of

$$
\Big(\prod_{a\in A'}x_a\Big)^{2m+n}\Big(\prod_{b\in B'}x_b^{-1}\Big)^{2m+n} \qquad (3.23)
$$

in (3.18). (When we here say "extract the coefficient of (3.23)," we treat "m" as if it were a formal variable. That is, in the sum on the right-hand side of (3.18), terms must be expanded, from each term which results from the expansion we factor out the product (3.23), and, if whatever remains is independent of m, it contributes to the coefficient.) This coefficient is the subsum of the sum on the right-hand side of (3.18) consisting of the summands corresponding to subsets A and B of $[n]$ with $A'=A\backslash B$ and $B'=B\backslash A$ such that $|A|+|B|=n$. We note at this point that this implies that $C:=[n]\backslash(A'\cup B')$ must have even cardinality. We let A'' be the intersection $A''=A\cap B$, so that $A=A'\dot\cup A''$ and $B=B'\dot\cup A''$ (with $\dot\cup$ denoting disjoint union), and we denote the complement of A'' in C by $(A'')^c$. Since $|A|+|B|=n$, we have $|A''|=|(A'')^c|$. Using the above notation, then, after some manipulation, the above described subsum can be rewritten as

$$
\begin{aligned}
&\frac{1}{D_1(n)}(-1)^{n|B'|+\frac12|C|}V(A')V(B')V\left((A')^{-1}\right)V\left((B')^{-1}\right)\\
&\quad\times R\left(A',(B')^{-1}\right)R\left(B',(A')^{-1}\right)R(C,A')\,R\left(C,(A')^{-1}\right)R\left(C^{-1},B'\right)R\left(C^{-1},(B')^{-1}\right)\\
&\quad\times\sum_{\substack{A''\subseteq C\\|A''|=\frac12|C|}}V(A'')V\left((A'')^{-1}\right)R\left(A'',(A'')^{-1}\right)V\left((A'')^c\right)V\left(((A'')^c)^{-1}\right)R\left((A'')^c,((A'')^c)^{-1}\right),
\end{aligned}
\qquad (3.24)
$$

where we used variations of the resultant notation $R(\dots)$, namely

$$
R(A,B):=\prod_{a\in A}\prod_{b\in B}(x_a-x_b),
$$

$$
R\left(A^{-1},B\right):=\prod_{a\in A}\prod_{b\in B}(x_a^{-1}-x_b),
$$

and

$$
R\left(A^{-1},B^{-1}\right):=\prod_{a\in A}\prod_{b\in B}(x_a^{-1}-x_b^{-1}).
$$

Now we turn our attention to (3.22). First of all, we observe that by exchanging the roles of A and B in the summand on the right-hand side of (3.22), the summand does not change except for a sign of $(-1)^{|A|-|B|}$. Thus, all summands corresponding to sets A and B whose cardinalities do not have the same parity cancel each other. We can therefore restrict the sum in (3.22) to subsets A and B of $[n]$ with $|A| \equiv |B| \pmod 2$. Consequently, if we extract the coefficient of (3.23) in (3.22), then, after some manipulation, we obtain

$$\frac{1}{D_1(n)}(-1)^{mn+n|B'|+\frac{1}{2}|C|}V(A')V(B')V\big((A')^{-1}\big)V\big((B')^{-1}\big)$$
$$\times R\big(A',(B')^{-1}\big)R\big(B',(A')^{-1}\big)R\big(C,A'\big)R\big(C,(A')^{-1}\big)R\big(C^{-1},B'\big)R\big(C^{-1},(B')^{-1}\big)$$
$$\times \sum_{A''\subseteq C} V(A'')V\big((A'')^{-1}\big)R\big(A'',((A'')^c)^{-1}\big)V\big((A'')^c\big)V\big(((A'')^c)^{-1}\big)R\big((A'')^c,(A'')^{-1}\big). \tag{3.25}$$

Here, with A and B as in (3.22), the meanings of A', B', A'', and C are $A' = A\cap B$, $B' = [n]\backslash(A\cup B)$, $A'' = A\backslash B$, $C = [n]\backslash A'\cup B'$, and $(A'')^c$ is the complement of A'' in C, so that $A = A'\dot\cup A''$ and $B = A'\dot\cup(A'')^c$. Because of the restriction $|A| \equiv |B| \pmod 2$, again C must have even cardinality. Clearly, clearing the factors common to (3.24) and (3.25), the equality of (3.24) and (3.25) is a direct consequence of (3.13). This completes the proof of the theorem.

Proof of Theorem 3.2. As we have shown when we rewrote Theorems 3.1 and 3.2 uniformly as (3.5), Theorem 3.2 results from the proof of Theorem 3.1 by "replacing m by $m+\frac{1}{2}$." Since, in the proof of Theorem 3.1, it is nowhere used that m is an integer, and since, in fact, m is treated there as a formal variable, Theorem 3.2 follows immediately.

3.5 Combinatorial interpretations

The purpose of this section is to give combinatorial interpretations of Theorems 3.1 and 3.2. As we already said in the introduction, these interpretations were at the origin of this work, which suggested Theorems 3.1 and 3.2 in the first place. They involve *rhombus tilings*, respectively *plane partitions*. When we speak of a rhombus tiling of some region, we always mean a tiling of the region by unit rhombi with angles of 60° and 120°. Examples of such tilings can be found in Figures 3.1–3.5. (Dotted lines and shadings should be ignored at this point.) We shall not recall the relevant plane partition definitions here, but instead refer the reader to [2], and to [6] for explanations on the relation between plane partitions and rhombus tilings of hexagons.

We claim that, by specializing $x_1 = x_2 = \cdots = x_n = 1$ in (3.1), we obtain the combinatorial factorization

$$PP(2m,n,n) = SPP(2m,n,n)\cdot TCPP(2m,n,n), \tag{3.26}$$

where $PP(2m,n,n)$ denotes the number of plane partitions contained in the $(2m)\times n\times n$ box (or, equivalently, the number of *rhombus tilings* of a hexagon with side lengths $2m,n,n,2m,n,n$; see Figure 3.1 for an example in which $m=2$ and $n=3$), $SPP(2m,n,n)$ denotes the number of *symmetric plane partitions* contained in the $(2m)\times n\times n$ box (or, equivalently, the number of rhombus tilings of a hexagon with side lengths $2m,n,n,2m,n,n$ which are symmetric with respect to the vertical symmetry axis of the hexagon; see Figure 3.2 for an example in which $m=2$ and $n=3$), and $TCPP(2m,n,n)$ denotes the number of *transpose complementary plane partitions* contained in the $(2m)\times n\times n$ box (or, equivalently, the number of rhombus tilings of a hexagon with side lengths $2m,n,n,2m,n,n$ which are symmetric with respect to the horizontal symmetry axis of the hexagon; see Figure 3.3.a for an example in which $m=n=3$; the dotted lines should be ignored for the moment).[1]

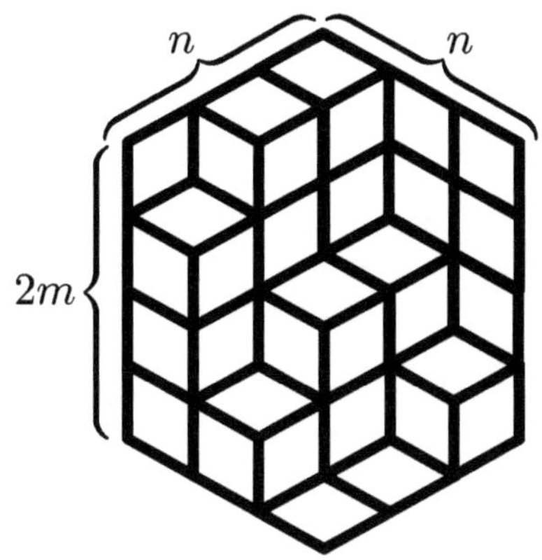

Fig. 3.1 A rhombus tiling of a hexagon

We start with the left-hand side of (3.26). The fact that the specialized Schur function $s_{((2m)^n)}(1,1,\dots,1)$ (with $2n$ occurrences of 1 in the argument) is equal to the number of plane partitions in the $(2m)\times n\times n$ box, and that this is equal to the number of rhombus tilings of a hexagon with side lengths $2m,n,n,2m,n,n$, is well-known (cf. [6] and [10, Sec. 7.21]).

It is also well-known (see [2, Sec. 4.3] or [8, Ch. I, Sec. 5, Ex. 15–17]) that the number of symmetric plane partitions contained in the $(2m)\times n\times n$ box is equal to $so_{(m^n)}(1,1,\dots,1)$ (with n occurrences of 1 in the argument).

The argument which explains that $so_{(m^n)}(-1,-1,\dots,-1)$ (with n occurrences of -1 in the argument) counts transpose complementary plane partitions contained in the $(2m)\times n\times n$ box (up to sign) is more elaborate. We begin by setting $x_i=-q^{i-1}$ in (3.7) with $N=n$ and $\lambda=(m^n)$, to obtain

$$so_{(m^n)}(-1,-q,-q^2,\dots,-q^{n-1})=(-1)^{mn}\frac{\det\limits_{1\le h,t\le n}\left(q^{(h-1)(m+n-t+\frac{1}{2})}+q^{-(h-1)(m+n-t+\frac{1}{2})}\right)}{\det\limits_{1\le h,t\le n}\left(q^{(h-1)(n-t+\frac{1}{2})}+q^{-(h-1)(n-t+\frac{1}{2})}\right)}.$$

[1] Clearly, the factorization (3.26) could be readily verified directly by using the known product formulas for $PP(2m,n,n)$, $SPP(2m,n,n)$, and $TCPP(2m,n,n)$ (cf. [2]). However, the point here is that it is a consequence of the more general factorization (3.1) featuring a Schur function and odd orthogonal characters.

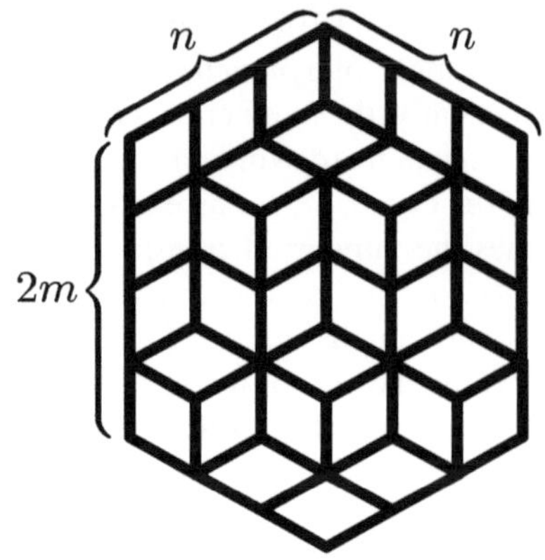

Fig. 3.2 A vertically symmetric rhombus tiling of a hexagon

The determinants on the right-hand side can be evaluated by means of (3.12). (This is seen by replacing h by $n+1-h$ and then interchanging the roles of h and t in the determinants above.) If we subsequently let q tend to 1, then we obtain

$$so_{(m^n)}(-1,-1,\dots,-1) = (-1)^{mn} \prod_{1\le h<t\le n} \frac{2m+2n+1-h-t}{2n+1-h-t}.$$

On the other hand, the product on the right-hand side is equal to a specialized symplectic character. Namely, by specializing $\lambda = (m^N)$, $x_i = q^i$, $i = 1,2,\dots,N$, in (3.9), using the identity (3.10), and finally letting q tend to 1, it is seen that

$$sp_{(m^N)}(1,1,\dots,1) = \prod_{1\le h<t\le N+1} \frac{2m+2N+3-h-t}{2N+3-h-t}$$

(with N occurrences of 1 in the argument). Thus, for $N = n-1$, we obtain that

$$so_{(m^n)}(-1,-1,\dots,-1) = (-1)^{mn}\, sp_{(m^{n-1})}(1,1,\dots,1).$$

To conclude the argument, it is easy to see using families of m lattice paths starting at the mid-points of the top-most m edges along the left vertical side of length $2m$ of the hexagon with side lengths $2m,n,n,2m,n,n$ and ending at the mid-points of the top-most m edges along the right vertical side of length $2m$ of the hexagon, the paths "following the rhombi" of the tiling (and, thus, staying necessarily in the upper half of the hexagon; see Figure 3.3.a,b), that rhombus tilings of a hexagon with side lengths $2m,n,n,2m,n,n$ which are symmetric with respect to the horizontal symmetry axis of the hexagon are in bijection with families $(P_1,P_2,\dots,P_m)$ of non-intersecting lattice paths consisting of horizontal and vertical unit steps, the path P_i starting at $(-i,i-1)$ and ending at $(n-i,n+i-1)$, $i = 1,2,\dots,m$, all paths never passing below the line $y = x+1$ (see Figure 3.3.c for the path family resulting from the one in Figure 3.3.b by deforming the paths to orthogonal ones;

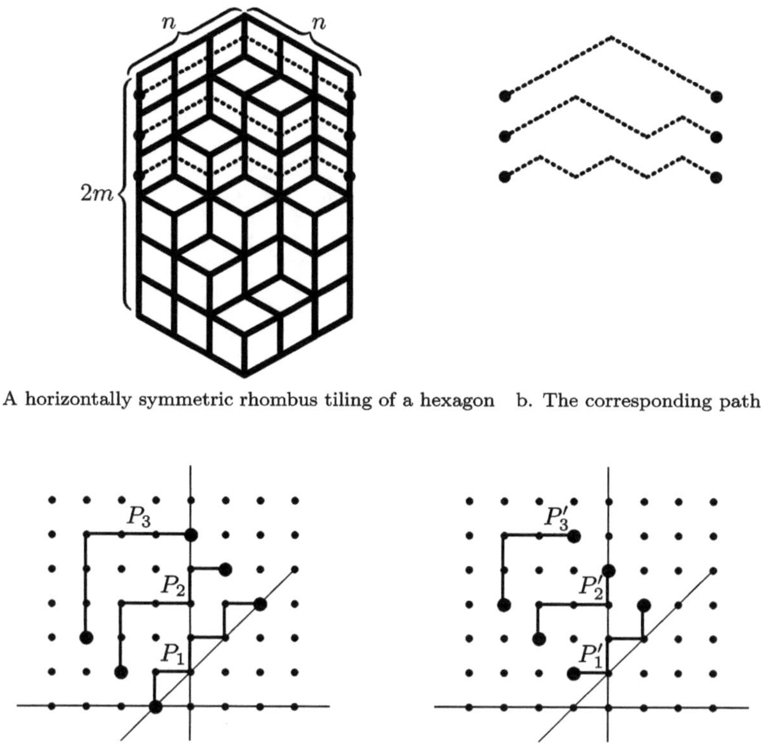

a. A horizontally symmetric rhombus tiling of a hexagon b. The corresponding path family

c. Orthogonal lattice paths d. Shorter non-intersecting lattice paths

Fig. 3.3

as usual, the term "non-intersecting" means that no two paths in the family have a common point). Because of the boundary $y = x + 1$ and the condition that paths are non-intersecting, the initial step of *any* path in such a family must be a vertical step while the final step of *any* path must be a horizontal step. Thus, our rhombus tilings are in bijection with families $(P_1, P_2, \dots, P_m)$ of non-intersecting lattice paths consisting of horizontal and vertical unit steps, the path P_i starting at $(-i, i)$ and ending at $(n-1-i, n-1+i)$, $i = 1, 2, \dots, m$, all paths never passing below the line $y = x + 1$ (see Figure 3.3.d). It is known (see [3, Sec. 5]) that these families of non-intersecting lattice paths are counted by $sp_{(m^{n-1})}(1, 1, \dots, 1)$ (with n occurrences of 1 in the argument). Thus, $|so_{(m^n)}(-1, -1, \dots, -1)|$ counts indeed rhombus tilings of a hexagon with side lengths $2m, n, n, 2m, n, n$ which are symmetric with respect to the horizontal symmetry axis of the hexagon, respectively transpose complementary plane partitions in the $(2m) \times n \times n$ box.

For Theorem 3.2 we are also able to provide a combinatorial interpretation in the context of rhombus tilings. However, it may be less convincing.

We specialize $x_1 = x_2 = \dots = x_n = 1$ in (3.2). Clearly, $s_{((2m+1)^n)}(1, 1, \dots, 1)$ (with $2n$ occurrences of 1 in the argument) counts plane partitions in the $(2m+1) \times n \times n$

box, respectively rhombus tilings of a hexagon with side lengths $2m+1,n,n,2m+1,n,n$.

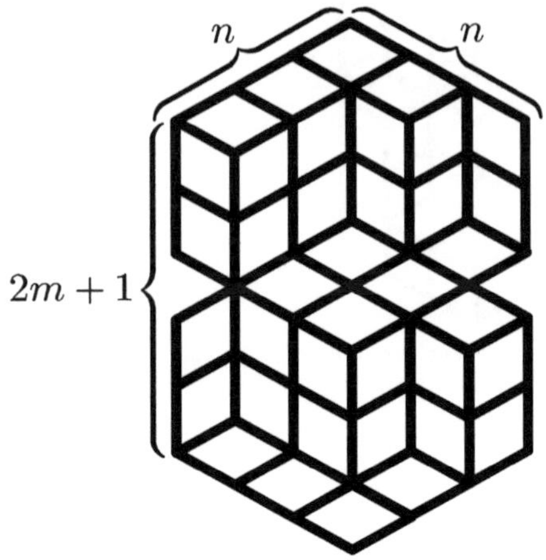

Fig. 3.4 A horizontally symmetric rhombus tiling of a hexagon with two missing triangles

Moreover, the argument above shows that $sp_{(m^n)}(1,1,\dots,1)$ (with n occurrences of 1 in the argument) counts rhombus tilings of a hexagon with side lengths $2m+1,n,n,2m+1,n,n$ from which two unit triangles have been removed on the left and the right end of the horizontal symmetry axis of the hexagon, the tilings being symmetric with respect to this axis (see Figure 3.4 for an example in which $m=2$ and $n=3$). Alternatively, this is also the number of horizontally symmetric rhombus tilings of a (full) hexagon with side lengths $2m,n+1,n+1,2m,n+1,n+1$ (compare with Figure 3.3.a).

In order to find a combinatorial interpretation of $o^{even}_{((m+1)^n)}(1,1,\dots,1)$ (with n occurrences of 1 in the argument), we start with the decomposition

$$o^{even}_{((m+1)^n)}(1,1,\dots,1) = so^{even}_{((m+1)^n)}(1,1,\dots,1) + so^{even}_{((m+1)^{n-1},-m-1)}(1,1,\dots,1),$$

where

$$so^{even}_{\lambda}(x_1,x_2,\dots,x_n) = \frac{\det\limits_{1\le h,t\le n}\left(x_h^{\lambda_t+n-t}+x_h^{-(\lambda_t+n-t)}\right)+\det\limits_{1\le h,t\le n}\left(x_h^{\lambda_t+n-t}-x_h^{-(\lambda_t+n-t)}\right)}{\det\limits_{1\le h,t\le n}\left(x_h^{n-t}+x_h^{-(n-t)}\right)}$$

is an irreducible character of $SO_{2n}(\mathbb{C})$ (and its spin covering group; see [4, (24.40)]). Here, $\lambda=(\lambda_1,\lambda_2,\dots,\lambda_n)$ is a non-increasing sequence of (possibly negative) integers or half-integers with $\lambda_{n-1}\ge|\lambda_n|$. It was proved in [1] (see [5] for a common generalization) that, for all non-negative integers or half-integers c, we have

$$so^{even}_{(c^n)}(x_1,x_2,\dots,x_n) = (x_1x_2\cdots x_n)^{-c}\cdot \sum_{\substack{\nu\subseteq((2c)^n)\\ \operatorname{oddcols}\left(((2c)^n)/\nu\right)=0}} s_\nu(x_1,x_2,\dots,x_n) \tag{3.27}$$

and

$$so^{even}_{(c^{n-1},-c)}(x_1,x_2,\dots,x_n) = (x_1x_2\cdots x_n)^{-c}\cdot \sum_{\substack{\nu\subseteq((2c)^n)\\ \operatorname{oddcols}\left(((2c)^n)/\nu\right)=2c}} s_\nu(x_1,x_2,\dots,x_n), \tag{3.28}$$

where $\operatorname{oddcols}\left(((2c)^n)/\nu\right)$ denotes the number of odd columns of the *skew diagram* (cf. [8, p. 4]) $((2c)^n)/\nu$. Thus, we have

$$o^{even}_{((m+1)^n)}(1,1,\dots,1) = \sum_{\nu\subseteq((2m+2)^n)}{}' \, s_\nu(1,1,\dots,1),$$

where $\sum'$ is taken over all diagrams ν with the property that $((2m+2)^n)/\nu$ consists either of only even columns or of only odd columns. (Both the even orthogonal character on the left-hand side and the Schur functions on the right-hand side contain n occurrences of 1 in their arguments.)

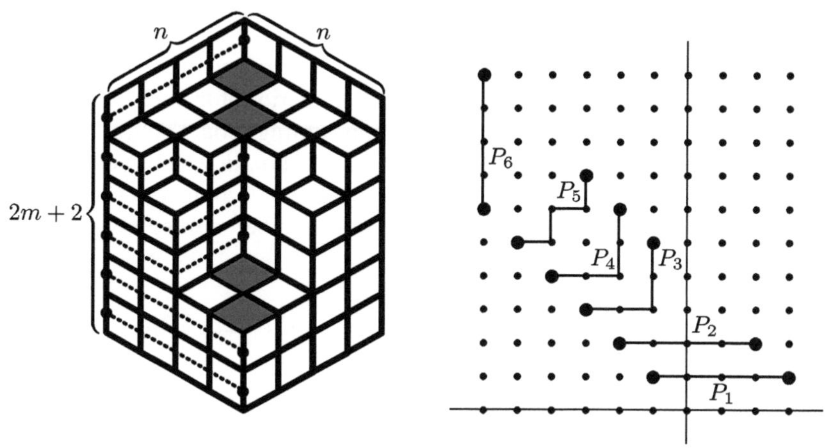

a. A vertically symmetric rhombus tiling of a hexagon b. Non-intersecting lattice paths

Fig. 3.5

Given a shape ν contained in $((2m+2)^n)$, it is well-known (cf. [3, Sec. 4] or [9, Ch. 4]) that $s_\nu(1,1,\dots,1)$ counts families $(P_1,\dots,P_{2m+2})$ of non-intersecting lattice paths consisting of horizontal and vertical unit steps, where the path P_i

runs from $(-i,i)$ to $(\nu'_i - i, n - \nu'_i + i)$, $i = 1,2,\dots,2m+2$. (Here, ν' denotes the partition *conjugate* to ν; cf. [8, p. 2]). On the other hand, in a similar way as above, such families of non-intersecting lattice paths (if ν is allowed to be any partition) are in bijection with rhombus tilings of a hexagon with side lengths $2m+2,n,n,2m+2,n,n$ which are symmetric with respect to the vertical symmetry axis of the hexagon (see Figure 3.5 for an example in which $m=2$, $n=4$, $\nu = (5,5,2,2)$). The property that $((2m+2)^n)/\nu$ consists either of only even columns or of only odd columns translates into the property that, in the corresponding rhombus tilings, chains of successive horizontally oriented rhombi along the vertical symmetry axis of the hexagon can appear in the interior of the hexagon only if they have even length. (By definition, a "chain of horizontally oriented rhombi along the vertical symmetry axis of the hexagon" is a set of horizontally oriented rhombi sitting on the vertical symmetry axis which, together, form a topologically connected set. For a chain, to be in the interior, means that none of the rhombi of the chain touches the boundary of the hexagon. In Figure 3.5.a, there are two such chains. They consist of the two contiguous strings of grey shaded rhombi, respectively.)

In summary, Theorem 3.2, when specialized to $x_1 = x_2 = \cdots = x_n = 1$, can be interpreted combinatorially as follows: the term $s_{((2m+1)^n)}(1,1,\dots,1)$ on the left-hand side counts rhombus tilings of a hexagon with side lengths $2m+1,n,n,2m+1,n,n$. The term $sp_{(m^n)}(1,1,\dots,1)$ counts horizontally symmetric rhombus tilings of a hexagon with side lengths $2m+1,n,n,2m+1,n,n$ from which two unit triangles have been removed on the left and the right end of the horizontal symmetry axis of the hexagon. Finally, the term $o^{even}_{((m+1)^n)}(1,1,\dots,1)$ counts vertically symmetric rhombus tilings of a hexagon with side lengths $2m+2,n,n,2m+2,n,n$ with the property that chains of successive horizontally oriented rhombi along the vertical symmetry axis of the hexagon can appear in the interior of the hexagon only if they have even length. We remark that these rhombus tilings could equivalently be seen as symmetric plane partitions in a $(2m+2)\times n\times n$ box in which "central terraces" (that is, horizontal levels situated along the plane of symmetry) must have even length, except at height 0 and at height $2m+2$. In other words, this specialization yields the identity

$$PP(2m+1,n,n) = TCPP(2m,n+1,n+1)\,SPP^*(2m+2,n,n), \tag{3.29}$$

where the * indicates that only those symmetric plane partitions are counted which satisfy the "even central terraces in the interior" condition described above. In particular, by specializing $x_h = q^{h-1}$ in (3.11), using (3.12) for evaluating the two determinants in (3.11), and then letting q tend to 1, we obtain that

$$SPP^*(2m,n,n) = 2\prod_{1\le h<t\le n} \frac{2m+2n-h-t}{2n-h-t}. \tag{3.30}$$

3.6 More factorization theorems

In this final section, we present two further factorization theorems similar to those of Theorems 3.1 and 3.2, which have been hinted at to us by Ron King. Since these theorems can be proved in a manner very similar to the one which led to our proofs of Theorems 3.1 and 3.2, we content ourselves with giving only sketches of proofs, leaving the (easily filled in) details to the reader.

Theorem 3.3. *For any non-negative integers m and n, we have*

$$s_{((2m+1)^n)}(x_1,x_1^{-1},x_2,x_2^{-1},\dots,x_n,x_n^{-1})+s_{((2m+1)^{n-1})}(x_1,x_1^{-1},x_2,x_2^{-1},\dots,x_n,x_n^{-1})\\ =(-1)^{mn}\,so_{((m+1)^n)}(x_1,x_2,\dots,x_n)\,so_{(m^n)}(-x_1,-x_2,\dots,-x_n). \quad (3.31)$$

Sketch of proof. We proceed in the same way as in the proof of Theorem 3.1. By comparison with (3.18), we see that the left-hand side of (3.31) is equal to

$$\frac{1}{D_1(n)}\sum_{\substack{A,B\subseteq[n]\\|A|+|B|=n}}(-1)^{\left(\sum_{a\in A}a\right)+\left(\sum_{b\in B}b\right)+n|B|-\binom{n+1}{2}}\left(\prod_{a\in A}x_a\right)^{2m+n+1}\left(\prod_{b\in B}x_b^{-1}\right)^{2m+n+1}\\ \cdot V(A)V(B^{-1})R(A,B^{-1})V(A^c)V\left((B^c)^{-1}\right)R\left(A^c,(B^c)^{-1}\right)\\ +\frac{1}{D_1(n)}\sum_{\substack{A,B\subseteq[n]\\|A|+|B|=n-1}}(-1)^{\left(\sum_{a\in A}a\right)+\left(\sum_{b\in B}b\right)+n|B|-\binom{n+1}{2}}\left(\prod_{a\in A}x_a\right)^{2m+n+2}\left(\prod_{b\in B}x_b^{-1}\right)^{2m+n+2}\\ \cdot V(A)V(B^{-1})R(A,B^{-1})V(A^c)V\left((B^c)^{-1}\right)R\left(A^c,(B^c)^{-1}\right), \quad (3.32)$$

with $D_1(n)$ having the same meaning as in the proof of Theorem 3.1. On the other hand, from (3.21) we see that the right-hand side of (3.31) is equal to

$$\frac{(-1)^{mn}}{D_1(n)}\sum_{A,B\subseteq[n]}(-1)^{\left(\sum_{a\in A}a\right)+\left(\sum_{b\in B}b\right)-\binom{|A|+1}{2}-\binom{|B|}{2}}\\ \cdot\left(\prod_{a\in A}x_a\right)^{m+\frac12}\left(\prod_{a\in A^c}x_a^{-1}\right)^{m+\frac12}\left(\prod_{b\in B}x_b\right)^{m+\frac32}\left(\prod_{b\in B^c}x_b^{-1}\right)^{m+\frac32}\\ \cdot V(A)V\left((A^c)^{-1}\right)R\left(A,(A^c)^{-1}\right)V(B)V\left((B^c)^{-1}\right)R\left(B,(B^c)^{-1}\right)\\ =\frac{(-1)^{mn}}{D_1(n)}\sum_{A,B\subseteq[n]}(-1)^{\left(\sum_{a\in A}a\right)+\left(\sum_{b\in B}b\right)-\binom{|A|+1}{2}-\binom{|B|}{2}}\\ \cdot\left(\prod_{a\in A\cap B}x_a\right)^{2m+2}\left(\prod_{a\in A^c\cap B^c}x_a^{-1}\right)^{2m+2}\left(\prod_{a\in A\cap B^c}x_b\right)^{-1}\left(\prod_{b\in A^c\cap B}x_b\right)\\ \cdot V(A)V\left((A^c)^{-1}\right)R\left(A,(A^c)^{-1}\right)V(B)V\left((B^c)^{-1}\right)R\left(B,(B^c)^{-1}\right). \quad (3.33)$$

We now fix disjoint subsets A' and B' of $[n]$ and extract the coefficients of

$$\left(\prod_{a\in A'}x_a\right)^{2m+n}\left(\prod_{b\in B'}x_b^{-1}\right)^{2m+n}$$

in (3.32) respectively in (3.33) (in the same sense as in Section 3.4). If $|A'|+|B'|$ has the same parity as n, then in (3.32) only the first sum contributes, while in (3.33) it is only terms where the cardinality of the symmetric difference $A\triangle B$ is even. One can then see in the same way as Theorem 3.1 followed from Lemma 3.1, that the equality of corresponding contributions follows from Lemma 3.2. On the other hand, if $|A'|+|B'|$ has parity different from the parity of n, then in (3.32) only the second sum contributes, while in (3.33) it is only terms where the cardinality of the symmetric difference $A\triangle B$ is odd. In this case, the corresponding equality follows from Lemma 3.3.

Theorem 3.4. *For any non-negative integers m and n, we have*

$$s_{((2m)^n)}(x_1,x_1^{-1},x_2,x_2^{-1},\dots,x_n,x_n^{-1})+s_{((2m)^{n-1})}(x_1,x_1^{-1},x_2,x_2^{-1},\dots,x_n,x_n^{-1})$$
$$=sp_{(m^n)}(x_1,x_2,\dots,x_n)\,o^{even}_{(m^n)}(x_1,x_2,\dots,x_n). \quad (3.34)$$

Sketch of proof. In a similar manner as we saw that the proof of Theorem 3.2 follows by "replacing m by $m+\frac{1}{2}$" in the proof of Theorem 3.1, the proof of Theorem 3.4 follows by "replacing m by $m-\frac{1}{2}$" in the proof of Theorem 3.3.

Again, using (3.3) and (3.4), Theorems 3.3 and 3.4 allow for a uniform statement, namely as

$$\left(\prod_{i=1}^{n}(x_i^{1/2}+x_i^{-1/2})\right)\Big(s_{(M^n)}(x_1,x_1^{-1},x_2,x_2^{-1},\dots,x_n,x_n^{-1})$$
$$+s_{(M^{n-1})}(x_1,x_1^{-1},x_2,x_2^{-1},\dots,x_n,x_n^{-1})\Big)$$
$$=so_{\left((\frac{M+1}{2})^n\right)}(x_1,x_2,\dots,x_n)\,o^{even}_{\left((\frac{M}{2})^n\right)}(x_1,x_2,\dots,x_n). \quad (3.35)$$

In a similar vein as in Section 3.5, by specializing $x_i=1$, $i=1,2,\dots,n$, in Theorems 3.3 and 3.4, we are able to derive potentially interesting combinatorial interpretations. If we perform these specializations in (3.31), then, by the combinatorial facts explained in Section 3.5, we obtain the identity

$$PP(2m+1,n,n)+PP(2m+1,n-1,n+1)=SPP(2m+2,n,n)\cdot TCPP(2m,n,n), \quad (3.36)$$

while the same specialization in (3.34) yields

$$PP(2m,n,n)+PP(2m,n-1,n+1)=TCPP(2m,n+1,n+1)\cdot SPP^*(2m,n,n), \quad (3.37)$$

where $PP(A,B,C)$ denotes the number of plane partitions contained in the $A\times B\times C$ box (or, equivalently, the number of rhombus tilings of a hexagon with side lengths A,B,C,A,B,C, $SPP(M,N,N)$ and $TCPP(M,N,N)$ have the same meaning as in Section 3.5, and where the * in $SPP^*(2m,n,n)$ means that the symmetric plane partitions in consideration satisfy the "even central terraces in the interior" condition explained at the end of Section 3.5.

Clearly again, the factorizations (3.36) or (3.37) could be readily verified directly by using the product formulas for the combinatorial quantities involved. However, it would have been difficult to see that they exist without having first considered the factorization identities for classical group characters given by Theorems 3.3 and 3.4, respectively.

Acknowledgements

We are indebted to Ron King and Soichi Okada for extremely insightful comments on the first version of this manuscript, which helped to improve its contents considerably.

References

1. A. J. Bracken and H. S. Green, *Algebraic identities for parafermi statistics of given order*, Nuovo Cimento, **9A** (1972), 349–365.
2. D. M. Bressoud, *Proofs and confirmations — The story of the alternating sign matrix conjecture*, Cambridge University Press, Cambridge, 1999.
3. M. Fulmek and C. Krattenthaler, *Lattice path proofs for determinant formulas for symplectic and orthogonal characters*, J. Combin. Theory Ser. A **77** (1997), 3–50.
4. W. Fulton and J. Harris, *Representation Theory*, Springer–Verlag, New York, 1991.
5. C. Krattenthaler, *Identities for classical group characters of nearly rectangular shape*, J. Algebra **209** (1998), 1–64.
6. G. Kuperberg, *Symmetries of plane partitions and the permanent determinant method*, J. Combin. Theory Ser. A **68** (1994), 115–151.
7. A. Lascoux, *Symmetric functions and combinatorial operators on polynomials*, CBMS Regional Conference Series in Mathematics, vol. 99, Amer. Math. Soc., Providence, RI, 2003.
8. I. G. Macdonald, *Symmetric Functions and Hall Polynomials*, second edition, Oxford University Press, New York/London, 1995.
9. B. E. Sagan, *The symmetric group*, 2nd edition, Graduate Texts in Math., vol. 203, Springer–Verlag, New York, 2001.
10. R. P. Stanley, *Enumerative Combinatorics*, Vol. 2, Cambridge University Press, Cambridge, 1999.

Chapter 4
On multivariate Newton-like inequalities

Leonid Gurvits

Abstract We study multivariate entire functions and polynomials with non-negative coefficients. A class of **Strongly Log-Concave** entire functions, generalizing *Minkowski* volume polynomials, is introduced: an entire function f in m variables is called **Strongly Log-Concave** if the function $(\partial x_1)^{c_1} ...(\partial x_m)^{c_m} f$ is either zero or $\log((\partial x_1)^{c_1} ...(\partial x_m)^{c_m} f)$ is concave on R_+^m. We start with yet another point of view (of *propagation*) on the standard univariate (or homogeneous bivariate) **Newton Inequalities**. We prove analogues of the **Newton Inequalities** in the multivariate **Strongly Log-Concave** case. One of the corollaries of our new Newton-like inequalities is the fact that the support $supp(f)$ of a **Strongly Log-Concave** entire function f is pseudo-convex (D-convex in our notation). The proofs are based on a natural convex relaxation of the derivatives $Der_f(r_1, ..., r_m)$ of f at zero and on the lower bounds on $Der_f(r_1, ..., r_m)$, which generalize the **van der Waerden-Egorychev-Falikman** inequality for the permanent of doubly-stochastic matrices. A few open questions are posed in the final section.

4.1 Introduction

This paper is concerned with multivariate polynomials and entire functions with nonnegative real coefficients.(All Taylor's series in this paper are taken at zero.) We continue the research, initiated in the recent papers [9], [10], [11], [7], [12] by the present author, on "combinatorics and combinatorial applications hidden in certain homogeneous polynomials with non-negative coefficients." Essentially, the main goal here is understanding how far one can push the approach from the above mentioned papers. The following definition introduces the main notation of the paper.

Leonid Gurvits
Los Alamos National Laboratory, Los Alamos, NM, e-mail: gurvits@lanl.gov

I.S. Kotsireas, E.V. Zima (eds.), *Advances in Combinatorial Mathematics*,
DOI 10.1007/978-3-642-03562-3_4,

Definition 4.1.

(i) We denote by $Sim(n)$ the standard simplex in R^n:

$$Sim(n) = \{(a_1,...,a_n) : a_i \geq 0, 1 \leq i \leq n; \sum_{1 \leq i \leq n} a_i = 1.$$

(ii) We denote by $Pol_+(m,n)$ the convex cone of polynomials with nonnegative coefficients in m variables of total degree n; the corresponding convex cone of homogeneous polynomials is denoted as $Hom_+(m,n)$.
We denote by $Ent_+(m)$ the convex cone of entire functions on C^m with non-negative coefficients in Taylor's series.

(iii) An entire function $f \in Ent_+(m)$ is called **Strongly Log-Concave** if for all integer vectors $(c_1,...,c_m) \in Z_+^m$ the function $(\partial x_1)^{c_1}...(\partial x_m)^{c_m} f$ is either zero or $\log((\partial x_1)^{c_1}...(\partial x_m)^{c_m} f)$ is concave on R_+^m. A set of **Strongly Log-Concave** polynomials $p \in Pol_+(m,n)$ is denoted as $SLC(m,n)$ and a set of **Strongly Log-Concave** entire functions $f \in Ent_+(m)$ is denoted as $SLC(m)$.

(iv) A (discrete) subset $S \subset Z^m$ is called D-convex if

$$Conv(S) \cap Z^m = S,$$

where $Conv(S)$ is the convex hull of S and Z^m is the m-dimensional integer lattice.
A map $G : Z^m \to [-\infty, +\infty]$ is called D-concave if

$$G(\sum_{1 \leq i \leq k < \infty} a_i Y_i) \geq \sum_{1 \leq i \leq k < \infty} a_i G(Y_i)$$

for all sequences $(a_1,...,a_k) \in Sim(k)$ and all vectors $Y_1,...,Y_k \in Z^m$ such that $\sum_{1 \leq i \leq k < \infty} a_i Y_i \in Z^m$.
Our notion of D-convexity coincides with the notion of **pseudo-convexity** *from [3]. As the term "pseudo-convex" is already occupied in the complex analysis, we think that the term D-***convexity** *is more appropriate (and informative).*

(v) The support of an entire function

$$f(x_1,\dots,x_m) = \sum_{(r_1,...,r_m) \in Z_+^m} a_{r_1,...,r_m} \prod_{1 \leq i \leq m} x_i^{r_i} \tag{4.1}$$

is defined as $supp(f) = \{(r_1,\dots,r_m) : a_{r_1,...,r_m} \neq 0\}$.

(vi) For an entire function $f \in Ent_+(m)$ and an integer vector $R = (r_1,\dots,r_m) \in Z_+^m$ we define $Der_f(R) =: (\partial x_1)^{r_1} \dots (\partial x_m)^{r_m} f(0)$.
In the notation of (4.1), $Der_f(R) = a_{r_1,...,r_m} \prod_{1 \leq i \leq m} r_i!$

Example 4.1.

(i) First, we note that a homogeneous polynomial $p \in Hom_{+}(m,n)$ is log-concave on R_{+}^{m} if and only if the function $p^{\frac{1}{n}}$ is concave on R_{+}^{m}.

(ii) A natural class of **Strongly Log-Concave** homogeneous polynomials in the convex cone $Hom_{+}(m,n)$ consists of **H-Stable** polynomials: a polynomial $p \in Hom_{C}(m,n)$ is called **H-Stable** if $p(Z) \neq 0$ provided $Re(Z) > 0$. It is easy to show and is well known that if $p \in Hom_{C}(m,n)$ is **H-Stable** then the polynomial $\frac{p}{p(x_1,...,x_m)} \in Hom_{+}(m,n)$ for any positive real vector $(x_1,...,x_m)$ and $(\partial x_i)p$ is either zero or **H-Stable**. Consider an univariate polynomial $R(t) = \sum_{0 \leq i \leq k} a_i t^i, a_k \neq 0$ and the associated homogeneous polynomial $p \in Hom_{+}(2,n)$, $p(x,y) = \sum_{0 \leq i \leq k} a_i x^i y^{n-i}$.
Then p is **H-Stable** iff the roots of R are non-positive real numbers, which shows that **H-Stable** polynomials are **Strongly Log-Concave**.

(iii) Another, distinct from **H-Stable**, class of **Strongly Log-Concave** homogeneous polynomials in the convex cone $Hom_{+}(m,n)$ consists of Minkowski polynomials $Vol_n\left(\sum_{1 \leq i \leq m} x_i K_i\right)$, where Vol_n stands for the standard volume in R^n and $K_1,...,K_m$ are convex compact subsets of R^n. The **Strong Log-Concavity** of Minkowski polynomials is essentially equivalent to the famous **Alexandrov-Fenchel** inequalities [1] for the mixed volumes.

Remark 4.1. We note that **H-Stable** and Minkowski polynomials satisfy a seemingly stronger property: they are invariant with respect to the changes of variables $Y = AX$, where A is a rectangular matrix with **non-negative** entries and without zero rows. We don't know whether such invariance holds in the general **Strongly Log-Concave** case.

We are interested in the following natural question: *when the support $supp(f)$ of an entire function $f \in Ent_{+}(m)$ is D-convex?*. Clearly, $supp(f)$ is D-convex if, for instance, the map $\log(Der_f) : Z^m \rightarrow [-\infty, +\infty)$ is D-concave. This is the case for $f(x,y) = \sum_{1 \leq i \leq n} a_i x^i y^{n-i}, f \in Hom_{+}(2,n)$ such that the univariate polynomial $R(t) = \sum_{1 \leq i \leq n} a_i t^i$ has only real roots. Spelling out the definition of D-concavity gives us a reformulation of the famous Newton's inequalities. In the case of **Strongly Log-Concave** multivariate entire functions, the map $\log(Der_f)$ is not necessary D-concave. We introduce the following map

$$C_f(r_1,...,r_m) = \inf_{x_i > 0} \frac{f(x_1,...,x_m)}{\prod_{1 \leq i \leq m} (\frac{x_i}{r_i})^{r_i}}, (r_1,...,r_m) \in Z_{+}^{m} \tag{4.2}$$

It is easy to show that if $f \in Ent_{+}(m)$ and $\log(f)$ is concave on R_{+}^{m} then $\log(C_f)$ is D-concave.
Therefore, the D-convexity of the support $supp(f)$ would follow from the property

$$C_f(R) > 0 \Leftrightarrow Der_f(R) > 0. \tag{4.3}$$

We prove in this paper a sharp quantitative version of (4.3):

$$\prod_{1\le i\le m} \frac{r_i!}{r_i^{r_i}} C_f(R) \ge Der_f(R) \ge \exp\left(-(\sum_{1\le i\le m} r_i)\right) C_f(R). \tag{4.4}$$

The inequalities (4.4) (and their more refined versions) generalize the van der Waerden-Egorychev-Falikman lower bound on the permanent of doubly-stochastic matrices [5], [6] and are used in this paper to prove Newton-like inequalities for **Strongly Log-Concave** entire functions.

4.2 Univariate Newton-like Inequalities

4.2.1 Propagatable sequences (weights)

Definition 4.2. Let us define the following closed subset of R^{n+1} of log-concave sequences:

$$LC = \{(d_0, \ldots, d_n) : d_i \ge 0, 0 \le i \le n; d_i^2 \ge d_{i-1}d_{i+1}, 1 \le i \le n-1\}.$$

We also associate with a given positive vector $(c_0, \ldots, c_n)$ the weighted shift operator $Shift_{\mathbf{c}} : R^{n+1} \longrightarrow R^{n+1}$,

$$Shift_{\mathbf{c}}((x_0, \ldots, x_n)^T) = (c_0x_1, \ldots, c_{n-1}x_n, 0)^T.$$

If c is the vector of all ones, then $Shift_{\mathbf{c}} =: Shift$.
A positive finite sequence $(b_0, \ldots, b_n)$ is called **propagatable** if the following implication holds:

$$(p^{(0)}(0)b_0, \ldots, p^{(n)}(0)b_n) \in LC \Longrightarrow (p^{(0)}(t)b_0, \ldots, p^{(n)}(t)b_n) \in LC, t \ge 0, \tag{4.5}$$

where p is a polynomial of degree at most n.

Analogously, we define infinite **propagatable** sequences by considering infinite log-concave sequences and entire functions in (4.5).

Proposition 4.1. *Let $c_0, \ldots, c_{n-1}$ be a nonnegative sequence. Then $exp(t(Shift_{\mathbf{c}}))(LC) \subset LC$ for all $t \ge 0$ if and only if*

$$2c_i \ge c_{i+1} + c_{i-1}, 1 \le i \le n-2; 2c_{n-1} \ge c_{n-2}.$$

(In other words, the infinite sequence $(c_0, \ldots, c_{n-1}, 0, \ldots)$ is concave).

Proof. (i) *The "only if" part:* Consider the linear system of differential equations:

$$X'(t) = Shift_{\mathbf{c}} X(t) : X(0) = (1,1,\ldots,1), X(t) = (X_0(t),\ldots,X_n(t)).$$

Suppose that $exp(tShift_{\mathbf{c}})(LC) \subset LC$ for all $t \geq 0$, i.e $X(t) \in LC$ for all $t \geq 0$. Define the following smooth functions:

$$r_i(t) = (X_i(t))^2 - X_{i+1}(t)X_{i-1}(t), 1 \leq i \leq n-1.$$

It follows that $r_i(0) = 0$ and $r_i(t) \geq 0$ for all $t \geq 0$. Therefore $r_i'(0) \geq 0$. Thus

$$0 \leq r_i'(0) = 2c_i - c_{i+1} - c_{i-1}, 1 \leq i \leq n-2; 0 \leq r_{n-1}'(0) = 2c_{n-1} - c_{n-2}.$$

(ii) *The "if" part:* As $exp(A) = \lim_{n\to\infty}(I + \frac{A}{n})^n$, thus it is sufficient to prove that $(I + tShift_{\mathbf{c}})(LC) \subset LC$ for all $t \geq 0$, which is done by straightforward derivations.

□

Remark 4.2. The observation that $(I + Shift)(LC) \subset LC$ is probably well known; we have learned it from Julius Borcea.

Theorem 4.1. *Let $(b_0,\ldots,b_k)$ be a positive sequence. Define $c_i = \frac{b_i}{b_{i+1}}, 0 \leq i \leq k-1$. The sequence $(b_0,\ldots,b_k)$ is* **propagatable** *iff the infinite sequence $(c_0, \ldots, c_{k-1}, 0, \ldots)$ is concave.*

Proof. Define a vector function $Mom_b(t) = (b_0 p^{(0)}(t),\ldots,b_n p^{(n)}(t))$.
Clearly, $Mom_b(t)$ solves the following system of linear differential equations:

$$Mom_b(t)' = Shift_{\mathbf{c}}(Mom_b(t)).$$

Therefore $(b_0,\ldots,b_n)$ is **propagatable** iff
$exp(t(Shift_{\mathbf{c}}))(LC) \subset LC$ for all $t \geq 0$. The result now follows from Proposition 4.1.
□

The following result follows fairly directly from Theorem 4.1.

Corollary 4.1. *Let $(b_0,\ldots,b_k,\ldots)$ be a positive infinite sequence. Define $c_i = \frac{b_i}{b_{i+1}}$, for $0 \leq i < \infty$. The sequence $(b_0,\ldots,b_k,\ldots)$ is* **propagatable** *iff the infinite sequence $(c_0, \ldots, c_{k-1}, \ldots)$ is concave.*

Example 4.2. A polynomial $p(t) = \sum_{0 \leq i \leq k} a_i t^i$ with nonnegative coefficients is called n-Newton for $n \geq k$ if

$$d_i^2 \geq d_{i-1}d_{i+1} : 1 \leq i \leq k-1, d_i =: \frac{a_i}{\binom{n}{i}}. \tag{4.6}$$

Or, in other words, the vector $(p^{(0)}(0)b_0,\ldots,p^{(k)}(0)b_k) \in LC$, where $b_i = (n-i)!$.
As $c_i = \frac{b_i}{b_{i+1}} = n - i : 0 \leq i \leq k-1$ hence it follows from Theorem 4.1 that
$(p^{(0)}(t)b_0,\ldots,p^{(k)}(t)b_k) \in LC : t \geq 0$. Equivalently,

$$(p^{(i+1)}(t))^2 \geq \frac{n-i}{n-i-1} p(i)(t)p^{(i+2)}(t) : t \geq 0, i \leq k-2, \tag{4.7}$$

which means that the functions $\sqrt[n-i]{p^{(i)}} : 0 \leq i \leq k$ are concave on R_+.

Let $f \in Ent_+(1)$ be entire univariate function, $f(t) = \sum_{0 \leq i < \infty} a_i t^i$.
A natural generalization of the n-Newton property, i.e. when $n \to \infty$, is the log-concavity of the infinite sequence $f^{(0)}(0), ..., f^{(k)}(0), ...$ Corollary 4.1 proves that this property is equivalent to **Strong Log-Concavity** of f.
We collect the above observations in the following proposition.

Proposition 4.2.

(i) A polynomial p with nonnegative coefficients is n-Newton, where $n \geq deg(p)$, iff the functions $\sqrt[n-i]{p^{(i)}} : 0 \leq i \leq k$ are concave on R_+.

Let us n-homogenize the univariate polynomial p, i.e. put $R(x,y) = y^n p(\frac{x}{y})$. Then, $R \in Hom_+(2,n)$ and the functions $\sqrt[n-i]{p^{(i)}} : 0 \leq i \leq k$ are concave on R_+ if and only if the polynomial R is **Strongly Log-Concave***.*

(ii) An entire function $f \in Ent_+(1)$ is **Strongly Log-Concave** *iff the infinite sequence $f^{(0)}(0), ..., f^{(k)}(0), ...$ is log-concave.*

Remark 4.3. The standard Newton Inequalities correspond to the case $n = deg(p)$ and hold if, for instance, the roots of p are real. It was proved by G. C. Shephard in [23] that a polynomial p is n-Newton iff $p(t) = Vol_n(tK_1 + K_2)$ for some convex compact subsets(simplexes) $K_1, K_2 \subset R^n$. This remarkable result can be used (see [14] and [15]) for alternative short proofs of Proposition 4.2 and Liggett's convolution theorem, which states that pq is $m+n$-Newton provided that p is n-Newton and p is m-Newton.
The literature on univariate Newton Inequalities is vast; we refer the reader to the recent survey [20]. But the results presented here seem to be new, nothing of the kind is mentioned in [20].

4.3 Multivariate Case

The main upshot of Proposition 4.2 is that in the univariate case as well in the bivariate homogeneous case the following equivalence holds:

"f is **Strongly Log-Concave**" $\Longleftrightarrow$ "the map $\log(Der_f)$ is D-concave".

In the general multivariate case both implications fail.

Example 4.3.

(i) Consider the polynomial

$$p(x_1,\ldots,x_{2n}) = (x_1+x_2)(x_2+x_3)\ldots(x_{2n-1}+x_{2n})(x_{2n}+x_1).$$

Clearly, it is **H-Stable**. Consider three vectors: $R_0 = (1,\ldots,1)$, $R_1 = (2,0,2,\ldots,0,2)$, $R_2 = (0,2,\ldots,0,2)$; $2R_0 = R_1 + R_2$. By direct inspection, $Der_p(R_0) = 2$, $Der_p(R_1) = Der_p(R_2) = 2^n$. Which gives

$$\log(Der_p(\frac{1}{2}(R_1+R_2))) = \frac{1}{2}(\log(Der_p(R_1)) + \log(Der_p(R_2))) - (n-1)\log(2). \tag{4.8}$$

(ii) *Alexandrov-Fenchel Inequalities.*
Take a homogeneous **Strongly Log-Concave** polynomial $p \in Hom_+(m,n)$ and fix a non-negative integer vector $R = (r_1, r_2, ..., r_m), \sum_{1\le i\le m} r_i = m$. Define the following polynomial $q \in Hom_+(2, n - \sum_{3\le i\le m} r_i)$,

$$q(x_1,x_2) = (\partial x_3)^{r_3} ... (\partial x_m)^{r_m} p(x_1,x_2,0,...,0).$$

Then q is either zero or **Strongly Log-Concave**. This observation leads to the following inequalities: if both vectors

$$R_1 = (r_1+1, r_2-1, r_3, ..., r_m), R_2 = (r_1-1, r_2+1, r_3, ..., r_m)$$

are non-negative then

$$Der_p(R) = Der_p(\frac{1}{2}(R_1+R_2)) \geq (Der_p(R_1))^{\frac{1}{2}}(Der_p(R_2))^{\frac{1}{2}} \tag{4.9}$$

(iii) Consider $p \in Hom_+(4,4), p(x_1,x_2,x_3,x_4) = x_1x_2x_3x_4 + \frac{1}{4}((x_1x_2)^2 + (x_3x_4)^2)$. Here the map $\log(Der_f)$ is D-concave but the polynomial p is not log-concave on R_+^4.

We prove in this paper that in the general multivariate case if f is **Strongly Log-Concave** then the map $\log(Der_f)$ is "almost" D-concave.

4.3.1 Generalized van der Waerden-Egorychev-Falikman lower bounds

This section follows the recent inductive approach by the author [10].

Definition 4.3. For an entire function $f \in Ent_+(n)$ we define its **Capacity** as

$$Cap(f) = \inf_{x_i>0} \frac{p(x_1,...,x_n)}{\prod_{1\le i\le n} x_i} \tag{4.10}$$

We need the following elementary result:

Lemma 4.1. *Consider a function $f: R_+ \rightarrow R_+$ such that the derivative $f'(0)$ exists.*

(i) If $f^{\frac{1}{k}}$ is concave on R_+ for $k>1$ then $f'(0) \geq (\frac{k-1}{k})^{k-1} \inf_{t>0} \frac{f(t)}{t}$.

(ii) If f is log-concave on R_+ then $f'(0) \geq \frac{1}{e} \inf_{t>0} \frac{f(t)}{t}$.
If, additionally, the function f is analytic and $f'(0) = \frac{1}{e} \inf_{t>0} \frac{f(t)}{t}$ then $f(t) = exp(at), a>0$.

(iii) Let $R(t) = a_0 + ... + a_n t^n$ be a strongly log-concave on R_+ univariate polynomial with nonnegative coefficients:
$G(i)^2 \geq G(i-1)G(i+1) : 1 \leq i \leq n-1, G(i) = a_i i!$.
Then $f'(0) \geq L(n) \inf_{t>0} \frac{f(t)}{t}$, where $L(n) = (\inf_{t>0} \frac{exp_n(t)}{t})^{-1}$ and the truncated exponential is defined as $exp_n(t) = 1 + ... + \frac{1}{n!} t^n$. (Note that exp_n is strongly log-concave on R_+.)

Proof.

(i) If $f(0) = 0$ then, obviously, $f'(0) \geq \inf_{t>0} \frac{f(t)}{t}$. Therefore, we can assume that $f(0) = 1$. As $f^{\frac{1}{k}}$ is concave and non-negative on R_+ thus
$f(t) \leq (1 + \frac{f'(0)}{k} t)^k, t \geq 0$.
The standard calculus gives us for $l(t) = (1 + \frac{f'(0)}{k} t)^k$ that

$$\inf_{t>0} \frac{l(t)}{t} = f'(0)(g(k))^{-1}, g(k) = \left(\frac{k-1}{k}\right)^{k-1}.$$

As $\inf_{t>0} \frac{f(t)}{t} \leq \inf_{t>0} \frac{l(t)}{t}$, we deduce that $f'(0) \geq g(k) \inf_{t>0} \frac{f(t)}{t}$.

(ii) As in the proof above, we can assume that $f(0) = 1$. It follows from the log-concavity that $f(t) \leq exp(f'(0)t), t \geq 0$. It is easy to see that

$$\inf_{t>0} \frac{f(t)}{t} \leq \inf_{t>0} \frac{exp(f'(0)t)}{t} = f'(0) exp(1) = \frac{exp(f'(0)s)}{s}, s = (f'(0))^{-1}.$$

Therefore, $f'(0) \geq \frac{1}{e} \inf_{t>0} \frac{f(t)}{t}$.
If $f'(0) = \frac{1}{e} \inf_{t>0} \frac{f(t)}{t}$ then, using the log-concavity again, we get that $f(t) = exp(f'(0)t), 0 \leq t \leq s$. If f is analytic then $f(z) = exp(az), z \in C, a = f'(0) > 0$.

(iii) Again, assume WLOG that $R(0) = 1$. It follows then from the strong log-concavity that

$$R(t) \leq 1 + ... + \frac{1}{n!} t^n = exp_n(t), t \geq 0.$$

The rest of the proof is now as above.
□

Corollary 4.2. *Let $f \in Ent_+(n+1)$ and $g_n(x_1, ..., x_n) = (\partial x_{n+1}) p(x_1, ..., x_n, 0)$. If f is log-concave on R_+^{n+1} then*

$$Cap(g_n) \geq \frac{1}{e} Cap(f). \tag{4.11}$$

If $p \in Hom_+(n+1, n+1)$ *is log-concave on* R_+^{n+1} *then*

$$Cap(q_n) \geq g(n+1) Cap(p), \text{ where } g(k) =: \left(\frac{k-1}{k}\right)^{k-1}. \tag{4.12}$$

Proof. We need to prove that $(\partial x_{n+1})p(x_1,...,x_n,0) \geq \frac{1}{e} Cap(p) \prod_{1 \leq i \leq n} x_i$. Define an univariate log-concave entire function $R(t) = f(x_1,...,x_n,t)$.
Then $R(t) \geq Cap(p)t \prod_{1 \leq i \leq n} x_i : t \geq 0$ and $R'(0) = (\partial x_{n+1}) f(x_1,...,x_n,0)$.
It follows from the second item in Lemma 4.1 that

$$(\partial x_{n+1})p(x_1,...,x_n,0) \geq \frac{1}{e} Cap(p) \prod_{1 \leq i \leq n} x_i.$$

The inequality (4.12) is proved in the very same way, using the first item in Lemma 4.1 and the fact that if $p \in Hom_+(n+1, n+1)$ is log-concave on R_+^{n+1} then $p^{\frac{1}{n+1}}$ is also concave on R_+^{n+1}. □

We use below the following notation:

$$vdw(n) = \frac{n!}{n^n}.$$

Theorem 4.2.

(i) *Let* $f \in Ent_+(n)$ *be* **Strongly Log-Concave** *entire function in n variables. Then the following inequality holds:*

$$Cap(f) \geq \frac{\partial^n}{\partial x_1 \ldots \partial x_n} f(0) \geq \frac{1}{e^n} Cap(f) \tag{4.13}$$

Note that the right inequality in (4.13) becomes equality if $f = exp\left(\sum_{1 \leq i \leq n} a_i x_i\right)$ *where* $a_i > 0, 1 \leq i \leq n$.

(ii) *Let a homogeneous polynomial* $p \in Hom_+(n,n)$ *be* **Strongly Log-Concave**. *Then the next inequality holds:*

$$Cap(f) \geq \frac{\partial^n}{\partial x_1 \ldots \partial x_n} f(0) \geq vdw(n) Cap(p) \tag{4.14}$$

Note that the right inequality in (4.14) becomes equality if $p = (\sum_{1 \leq i \leq n} a_i x_i)^n$ *where* $a_i > 0, 1 \leq i \leq n$.

(iii) *Let a polynomial* $p \in Pol_+(n,n)$ *be* **Strongly Log-Concave**. *Then the next inequality holds:*

$$Cap(f) \geq \frac{\partial^n}{\partial x_1 ... \partial x_n} f(0) \geq \prod_{1 \leq i \leq n} L(i) Cap(p), \tag{4.15}$$

where $L(n) = (\inf_{t>0} \frac{exp_n(t)}{t})^{-1}$.
(Note that $L(1) = 1, L(2) = (1+\sqrt{2})^{-1}$ *and* $L(n) > e^{-1}, n \geq 1$.*)*

Proof.

(i) Define the following entire functions $q_i \in Ent_+(i)$: $q_n = f$, $q_i(x_1, \dots, x_i) = \frac{\partial^{n-i}}{\partial x_{i+1} ... \partial x_n} f(x_1, \dots, x_i, 0, \dots, 0)$. Notice that $q_1'(0) = \frac{\partial^n}{\partial x_1 ... \partial x_n} f(0)$. By the definition of **Strongly Log-Concavity**, these entire functions are either log-concave or zero. Using the inequality (4.11), we get that

$$Cap(q_i) \geq \frac{1}{e} Cap(q_{i+1}), 1 \leq i \leq n-1.$$

Therefore

$$\inf_{t>0} \frac{q_1(t)}{t} = Cap(q_1) \geq \left(\frac{1}{e}\right)^{n-1} Cap(f).$$

Finally, using Lemma 4.1, we get that

$$\frac{\partial^n}{\partial x_1 \dots \partial x_n} f(0) = q_1'(0) \geq \frac{1}{e} \inf_{t>0} \frac{q_1(t)}{t} \geq \frac{1}{e^n} Cap(f).$$

(ii) If a homogeneous polynomial $p \in Hom_+(n,n)$ is **Strongly Log-Concave** then the polynomials $q_i \in Hom_+(i,i)$, $\frac{\partial^n}{\partial x_1 ... \partial x_n} p(0) = Cap(q_1)$ and $(q_i)^{\frac{1}{i}}$ is concave on $R^i_+, 1 \leq i \leq n$. It follows from the inequality (4.12) that

$$\frac{\partial^n}{\partial x_1 \dots \partial x_n} p(0) = Cap(q_1) \geq \prod_{2 \leq k \leq n} g(k) Cap(p) = \frac{n!}{n^n} Cap(p).$$

□

4.3.2 General monomials

Consider an entire function $f \in Ent_+(m)$ and an integer non-negative vector $R = (r_1, ..., r_m)$. Assume WLOG that $R = (r_1, ..., r_k, 0, ..., 0) : r_i > 0, 1 \leq i \leq k; k \leq n$. Let us define the entire function $f_R \in Ent_+(|R|_1)$, where $|R|_1 = r_1 + ... + r_k$.

$$f_{(R)}(y_1, \dots, y_{|R|_1}) = f(e_1(y_1 + \dots + y_{r_1}) + \dots + e_k(y_{r_1 + ... + r_{k-1}+1} + \dots + y_{r_1 + ... + r_k})),$$

where $\{e_1, \dots, e_m\}$ is the standard basis in C^m. The following identity is obvious:

$$(\partial x_1)^{r_1} \dots (\partial x_m)^{r_m} f(0) = (\partial y_1) \dots (\partial y_{|R|_1}) f_{(R)}(0).$$

Note that if the original entire function (homogeneous polynomial) f is Strongly Log-Concave (H-Stable) then the same holds for the entire function (homogeneous polynomial) $f_{(R)}$.

It easily follows from the arithmetic-geometric means inequality that

$$Cap(f_{(R)}) = C_f(r_1,\dots,r_m) =: \inf_{x_i>0} \frac{f(x_1,\dots,x_m)}{\prod_{1\leq i\leq m}(\frac{x_i}{r_i})^{r_i}} \tag{4.16}$$

As we deal only with entire functions with the non-negative coefficients, the following inequality holds:

$$\left(\prod_{1\leq i\leq m} vdw(r_i)\right) C_f(r_1,\dots,r_m) \geq (\partial x_1)^{r_1}\dots(\partial x_m)^{r_m} f(0) \tag{4.17}$$

Putting these observations together, we get the Corollary to Theorem 4.2.

Corollary 4.3.

(i) *Let $f \in Ent_+(m)$ be* **Strongly Log-Concave** *entire function in m variables. Then for all integer vectors $R = (r_1,\dots,r_m) \in Z_+^m$ the next inequalities hold:*

$$\begin{aligned}\left(\prod_{1\leq i\leq m} vdw(r_i)\right) C_f(r_1,\dots,r_m) &\geq (\partial x_1)^{r_1}\dots(\partial x_m)^{r_m} f(0) \geq \\ &\geq exp(-|R|_1) C_f(r_1,\dots,r_m)\end{aligned} \tag{4.18}$$

(ii) *Let a homogeneous polynomial $p \in Hom_+(m,n)$ be* **Strongly Log-Concave**. *Then for all integer vectors $R = (r_1,\dots,r_m) \in Z_+^m, \sum_{1\leq i\leq m} r_i = n$ the next inequalities hold:*

$$\begin{aligned}\left(\prod_{1\leq i\leq m} vdw(r_i)\right) C_p(r_1,\dots,r_m) &\geq (\partial x_1)^{r_1}\dots(\partial x_m)^{r_m} p(0) \geq \\ &\geq vdw(n) C_p(r_1,\dots,r_m)\end{aligned} \tag{4.19}$$

Let us recall the generalized Schrijver's inequality from [10].

Theorem 4.3. *Let $p \in Hom_+(n,n)$ be* **H-Stable**. *Let us denote the degree of variable x_i in the polynomial p as $deg_p(i)$.*
If $deg_p(i) \leq k \leq n, 1 \leq i \leq n$, Then the next inequality holds:

$$Cap(p) \geq \frac{\partial^n}{\partial x_1 \dots \partial x_n} p(0) \geq \left(\frac{k-1}{k}\right)^{(k-1)(n-k)} vdw(k) Cap(p) \tag{4.20}$$

Combining Theorem 4.3 and observations (4.16), (4.17) we get the following:

Corollary 4.4. *Let $p \in Hom_+(n,n)$ be* **H-Stable**. *Assume that the degree of variable x_i in the polynomial p, $deg_p(i) \leq k \leq n, 1 \leq i \leq n$. Then the following inequalities hold:*

$$\left(\prod_{1\le i\le m} vdw(r_i)\right) C_p(r_1,\dots,r_m) \ge (\partial x_1)^{r_1}\dots(\partial x_m)^{r_m} p(0) \ge \\ \ge \left(\frac{k-1}{k}\right)^{(k-1)(n-k)} vdw(k) C_p(r_1,\dots,r_m) \tag{4.21}$$

4.3.3 *A lower bound on the inner products of* H-Stable *polynomials*

Theorem 4.4. *Let us consider two* **H-Stable** *polynomials $p,q \in Hom_+(m,n)$:*

$$p(x_1,\dots,x_m) = \sum_{r_1+\dots+r_m=n} a_{r_1,\dots,r_m} \prod_{1\le i\le m} x_i^{r_i},$$

$$q(x_1,\dots,x_m) = \sum_{r_1+\dots+r_m=n} b_{r_1,\dots,r_m} \prod_{1\le i\le m} x_i^{r_i},$$

and a nonnegative vector $(l_1,\dots,l_m)$ such that $\sum_{1\le i\le m} l_i = n$.

Let us assume that

$$\inf_{x_i>0, 1\le i\le m} \frac{p(x_1,\dots,x_m)}{\prod_{1\le i\le m} x_i^{l_i}} =: A > 0, \quad \inf_{x_i>0, 1\le i\le m} \frac{q(x_1,\dots,x_m)}{\prod_{1\le i\le m} x_i^{l_i}} =: B > 0. \tag{4.22}$$

Then the following inequality holds:

$$< p,g >=: \sum_{r_1+\dots+r_m=n} a_{r_1,\dots,r_m} b_{r_1,\dots,r_m} \ge AB\frac{vdw(nm)}{vdw(n)^m} \tag{4.23}$$

Proof. Let us consider a rational function

$$F = \prod_{1\le i\le m} x_i^n p(x_1,\dots,x_m) q(\frac{1}{x_1},\dots,\frac{1}{x_m}).$$

It is clear that, in fact, $F \in Hom_+(m,nm)$ and F is **H-Stable**. Note that

$$(n!)^m \sum_{r_1+\dots+r_m=n} a_{r_1,\dots,r_m} b_{r_1,\dots,r_m} = (\partial x_1)^n \dots (\partial x_m)^n F(0).$$

It follows from (4.22) that $C_F(n,\dots,n) \ge ABn^{nm}$. Using the right inequality in (4.19), we get that

$$\sum_{r_1+\dots+r_m=n} a_{r_1,\dots,r_m} b_{r_1,\dots,r_m} \ge AB\frac{vdw(nm)}{vdw(n)^m}.$$

□

Remark 4.4.

(i) It is easy to see that the inequalities (4.22) hold for some vector $(l_1,\ldots,l_m)$ if and only if the Newton polytopes, $Newt(p)$ and $Newt(q)$, have non-empty intersection. (Recall that the Newton polytope $Newt(p)$ is the convex hull of the support $supp(p)$.)
One of the corollaries of Theorem 4.4 is the fact that the intersection $Newt(p)\cap Newt(q)$ is not empty iff the intersection $supp(p)\cap supp(q)$ is not empty. There is alternative(and harder) way to prove this fact. It was proved in [9] and [11] that if p is a **H-Stable** polynomial then the Newton polytope $Newt(p)$ is the **polymatroid**, based on some integer valued submodular function. It follows from the celebrated Edmonds' result [4] that all the vertices of $Newt(p)\cap Newt(q)$ are integer. Therefore, if $Newt(p)\cap Newt(q)$ is not empty then there exists an integer vector $(r_1,\ldots,r_m)\in Newt(p)\cap Newt(q)$. But all integer vectors in $Newt(p)$ (resp. $Newt(q)$) belong to the support $supp(p)$ (resp. $supp(q)$).
The inequality (4.23) is unlikely sharp. We conjecture here a sharp version:

$$\sum_{r_1+\ldots+r_m=n} a_{r_1,\ldots,r_m} b_{r_1,\ldots,r_m} \prod_{1\le i\le m} (r_i)! \ge AB\frac{n!}{m^n}.$$

(ii) If **H-Stable** polynomials $p,q\in Hom_+(m,n)$ are both multi-linear, i.e. $deg_p(i)$, $deg_q(i)\le 1$, $1\le i\le m$, then $<p,q>=(\partial x_1)\ldots(\partial x_m)G(0)$, where

$$G(x_1,\ldots,x_m)=(\prod_{1\le i\le m} x_i)p(x_1,\ldots,x_m)q(\frac{1}{x_1},\ldots,\frac{1}{x_m}).$$

Note that the polynomial $G\in Hom_+(m,m)$ is **H-Stable**, $Cap(G)\ge AB$ and $deg_G(i)\le 2, 1\le i\le m$. Using Theorem 4.3, we get the following inequality:

$$<p,q>\ \ge\ AB2^{-m+1} \tag{4.24}$$

The inequality (4.24) is sharp for $m=2n$.

4.4 Multivariate Newton Inequalities

We start with the following simple fact.

Proposition 4.3. *If an entire function $f\in Ent_+(m)$ is log-concave on R_+^m then the map C_f, defined as*

$$C_f(y_1,\ldots,y_m)=\inf_{x_i>0}\frac{f(x_1,\ldots,x_m)}{\prod_{1\le i\le m}(\frac{x_i}{y_i})^{y_i}}, y_i\ge 0$$

is log-concave on R_+^m.

Proof. Assume WLOG that $y_i > 0, 1 \leq i \leq k \leq m$ and $y_j = 0, k+1 \leq j \leq m$. It follows from the monotonicity of f that

$$C_f(x_1,\dots,x_m) = \inf_{x_i>0, 1\leq i\leq k} \frac{f(x_1,\dots,x_k,0,\dots,0)}{\prod_{1\leq i\leq k}(\frac{x_i}{y_i})^{y_i}}.$$

Therefore

$$C_f(y_1,\dots,y_m) \geq a \text{ iff } \log(f(x_1y_1,\dots,x_my_m)) \geq \log(a) + \sum_{1\leq i\leq m} y_i \log(x_i)$$

for all positive vectors $(x_1,\dots,x_m)$. The desired log-concavity follows now from the log-concavity of the function f and of the logarithm. □

Let $Y = (r_1,\dots,r_m) \in Z_+^m$ be an integer vector. We use below the following notations:

$$VDW(Y) = \prod_{1\leq i\leq m} vdw(r_i), \text{ where } vdw((r) = \frac{r!}{r^r}.$$

Theorem 4.5. *Let us consider integer vectors $Y_0, Y_1, \dots, Y_k \in Z_+^m$ such that*

$$Y_0 = \sum_{1\leq i\leq k} a_i Y_i; a_i \geq 0, \sum_{1\leq i\leq k} a_i = 1.$$

(i) Suppose that the entire function $f \in Ent_+(m)$ is **Strongly Log-Concave**. *Then*

$$Der_f(Y_0) \geq \left(exp(-|Y_0|_1) \prod_{1\leq i\leq k} (VDW(Y_i))^{-a_i} \right) \prod_{1\leq i\leq k} (Der_f(Y_i))^{a_i} \quad (4.25)$$

(ii) If $p \in Hom_+(m,n)$ is **Strongly Log-Concave** *then*

$$Der_p(Y_0) \geq \left(vdw(n) \prod_{1\leq i\leq k} (VDW(Y_i))^{-a_i} \right) \prod_{1\leq i\leq k} (Der_p(Y_i))^{a_i} \quad (4.26)$$

(iii) If $p \in Hom_+(m,n)$ is **H-Stable** *and $deg_p(i) \leq k \leq n$ for all $1 \leq i \leq m$ then*

$$Der_p(Y_0) \geq$$
$$\geq \left(\left(\tfrac{k-1}{k}\right)^{(k-1)(n-k)} vdw(k) \prod_{1\leq i\leq k} (VDW(Y_i))^{-a_i} \right) \prod_{1\leq i\leq k} (Der_p(Y_i))^{a_i} \quad (4.27)$$

Proof. We will prove only the inequality (4.25) as the other ones are proved in the same way.
Using the the right inequality in (4.18), we get that

$$Der_f(Y_0) \geq exp(-|Y_0|_1) C_f(Y_0).$$

Since the map C_f is log-concave hence

$$C_f(Y_0) \geq \prod_{1 \leq i \leq k} (C_f(Y_i))^{a_i}.$$

Finally, we use the left inequality in (4.18):

$$C_f(Y_i) \geq (VDW(Y_i))^{-1} Der_f(Y_i).$$

□

Corollary 4.5. *The support $supp(f)$ of* **Strongly Log-Concave** *entire function $f \in Ent_+(m)$ is D-convex.*

Example 4.4.

(i) Let us consider the following vectors in Z_+^n:

$$Y_0 = (1,1,\ldots,1); Y_1 = (n,0,\ldots,0),\ldots,Y_n = (0,0,\ldots,n).$$

Note that $Y_0 = \sum_{1 \leq i \leq n} \frac{1}{n} Y_i$. If $p \in Hom_+(n,n)$ is **Strongly Log-Concave** then (4.26) gives the next inequality

$$Der_p(Y_0) \geq \prod_{1 \leq i \leq k} (Der_p(Y_i))^{\frac{1}{n}},$$

which is attained on $p(x_1,\ldots,x_n) = (x_1 + \ldots + x_n)^n$.

(ii) Consider three vectors in Z_+^{2n}:

$$Y_0 = (1,1,\ldots,1),\ Y_1 = (2,\ldots,2,0,\ldots,0),\ Y_2 = (0,\ldots,0,2,\ldots,2),$$

satisfying $|Y_1|_1 = |Y_2|_1 = 2n$.
If $p \in Hom_+(2n,2n)$ is **H-Stable** and $deg_p(i) \leq 2 \leq 2n$ for all $1 \leq i \leq 2n$ then it follows from (4.27) that

$$Der_p(Y_0) \geq 2^{-n+1} \prod_{1 \leq i \leq 2} (Der_p(Y_i))^{\frac{1}{2}}. \tag{4.28}$$

The inequality (4.28) is attained on the polynomial

$$p(x_1,\ldots,x_{2n}) = (x_1+x_2)(x_2+x_3)\ldots(x_{2n-1}+x_{2n})(x_{2n}+x_1).$$

4.5 Comments and open problems

(i) The inequality (4.14) is a far reaching generalization of the famous van der Waerden conjecture on the permanent of doubly-stochastic matrices ([19], [5], [6] and the Bapat's conjecture [2]), [12]. See more on this combinatorial connection in [11], [13], [7].

The van der Waerden conjecture corresponds to **H-Stable** polynomials

$$Prod_A(x_1,...,x_n) = \prod_{1\leq i\leq n}\prod_{1\leq j\leq n} A(i,j)x_j,$$

where $n\times n$ matrix is non-negative entry-wise and has no zero rows. If such a matrix is doubly-stochastic, i.e. all its rows and columns sum to 1, then $Cap(Prod_A) = 1$.
The convex relaxation approach to Newton-like inequalities in Theorem 4.5 was introduced by the author in [12] for the determinantal polynomials

$$\det\left(\sum_{1\leq i\leq m} x_i A_i\right),$$

where $A_1,\ldots,A_m$ are $n\times n$ hermitian PSD matrices. The corresponding inequalities in [12] are weaker than in the present paper.

(ii) The log-concavity of f alone is not sufficient for D-convexity of its support $supp(f)$ even for univariate polynomials with non-negative coefficients. Indeed, consider $p(t) = t+t^3$. The fourth root $\sqrt[4]{p(t)}$ is concave on R_+:

$$(p^{(1)}(t))^3 - \frac{4}{3}p(t)p^{(2)}(t) = (1+3t^2)^2 - \frac{4}{3}(t+t^3)6t = (t^2-1)^2 \geq 0.$$

This example can be lifted to a "bad" log-concave homogeneous polynomial $q\in Hom_+(4,4)$:
$q(x,y,v,w) = (x+y)^3(v+w)+(v+w)^3(x+y)$. It is easy to see that $Cap(q) = 2^5$ but

$$\frac{\partial^4}{\partial x\partial y\partial v\partial w}q(0) = 0.$$

(iii) In the case of **H-Stable** polynomials, Corollary 4.5 can be made much more precise:
Define, for a subset $S\subset\{1,...,m\}$ and a polynomial $p\in Hom_+(m,n)$, the integer number $Deg_p(S)$ equal to the maximum total degree attained on variables in S.
Then the following relation holds:

$$a_{r_1,...,r_m} > 0 \Longleftrightarrow \sum_{j\in S} r_j \leq Deg_p(S) : S\subset\{1,...,m\}, p\in Hom_+(m,n). \quad (4.29)$$

Additionally, the integer valued map $Deg_p : 2^{\{1,...,m\}} \to \{0,\ldots,n\}$ is **submodular**.
The characterization (4.29), proved in [9], is a far reaching generalization of the Hall-Rado theorems on the existence of perfect matchings.
The paper [11] provides algorithmic applications of this result: strongly polynomial deterministic algorithms for the membership problem as for the support as well for the Newton polytope of **H-Stable** polynomials $p\in Hom_+(m,n)$,

given as oracles.
We don't know whether (4.29) works for **Strongly Log-Concave** homogeneous polynomials. But it would follow from the following conjecture/question:

Conjecture 4.1. Let $p \in Hom_{+}(3,n)$ be **Strongly Log-Concave**. Then there exist convex compact subsets $K_1, K_2, K_3 \subset R^n$ such that

$$p(x_1,x_2,x_3) = Vol_n(x_1K_1 + x_2K_2 + x_3K_3) : x_1,x_2,x_3 \geq 0 \tag{4.30}$$

Or put more modestly:

Question 4.1. Which **Strongly Log-Concave** polynomials $p \in Hom_{+}(3,n)$ allow the representation (4.30)?

The Minkowski polynomials $Vol_n \in Hom_{+}(3,n), Vol_n(x_1K_1 + x_2K_2 + x_3K_3)$ actually have seemingly stronger, than **Strong Log-Concavity**, *property: the polynomials $\prod_{1 \leq j \leq r < n}(\sum_{1 \leq i \leq 3}(a_{i,j}\partial x_i)Vol_n$ are either zero or log-concave on R^3_+ provided that $a_{i,j} \geq 0$.*

(iv) *Question 4.2.* Is the set of **Strongly Log-Concave** entire functions closed under the multiplications?
If true it would imply that the Minkowski sum $supp(f) + supp(g) = supp(fg)$ is D-convex if f,g are **Strongly Log-Concave**.
We note that the problem of D-convexity of Minkowski sums of D-convex sets was studied from a combinatorial point of view in [3].
(v) *Question 4.3.* What are the asymptotically exact constants in Theorem 4.5? Are the cyclic polynomials in Example 4.3 extremal?
(vi) Can recently refuted Okounkov's conjecture [21], in the representation theory, on log-concavity of multiplicities be fixed/generalized in the way similar to Theorem 4.5?
(vii) Stable multivariate polynomials form a backbone of linear multivariate control. If $p \in HSP_{+}(n,n)$ then $Cap(p) = \inf_{Re(z_i)>0} \frac{|p(z_1,\ldots,z_n)|}{\prod_{1 \leq i \leq n} Re(z_i)}$. In other words, the capacity can be viewed as a measure of stability. What is a meaning of capacity if terms of control/dynamics or in terms of the corresponding hyperbolic PDE?
(viii) Can our results be reasonably generalized to the fractional derivatives?

References

1. A. Aleksandrov, On the theory of mixed volumes of convex bodies, IV, Mixed discriminants and mixed volumes (in Russian), *Mat. Sb. (N.S.)* 3 (1938), 227-251.
2. R.B. Bapat, Mixed discriminants of positive semidefinite matrices, *Linear Algebra and its Applications* 126, 107-124, 1989.

3. V.I. Danilov and G.A. Koshevoy, Discrete convexity and unimodularity, Advances in Mathematics, Volume 189, Issue 2, 20 December 2004, Pages 301-324.
4. J. Edmonds, Submodular functions, matroids, and certain polyhedra, in: R. Guy, H. Hanani, N. Sauer and J. Schonheim (eds.), *Combinatorial Structures and Their Applications*, Gordon and Breach, New York, 1970, 69-87.
5. G.P. Egorychev, The solution of van der Waerden's problem for permanents, *Advances in Math.*, 42, 299-305, 1981.
6. D.I. Falikman, Proof of the van der Waerden's conjecture on the permanent of a doubly stochastic matrix, *Mat. Zametki* 29, 6: 931-938, 957, 1981, (in Russian).
7. S. Friedland and L. Gurvits, Lower Bounds for Partial Matchings in Regular Bipartite Graphs and Applications to the Monomer-Dimer Entropy, *Combinatorics, Probability and Computing*, 2008.
8. L. Garding, An inequality for hyperbolic polynomials, Jour. of Math. and Mech., 8(6): 957-965, 1959.
9. L. Gurvits, Combinatorial and algorithmic aspects of hyperbolic polynomials, 2004; available at http://xxx.lanl.gov/abs/math.CO/0404474.
10. L. Gurvits, A proof of hyperbolic van der Waerden conjecture: the right generalization is the ultimate simplification, Electronic Colloquium on Computational Complexity (ECCC)(103): (2008) and arXiv:math/0504397.
11. L. Gurvits, Hyperbolic polynomials approach to van der Waerden/Schrijver-Valiant like conjectures: sharper bounds, simpler proofs and algorithmic applications, *Proc. 38 ACM Symp. on Theory of Computing (StOC-2006)*, 417-426, ACM, New York, 2006.
12. L. Gurvits, van der Waerden Conjecture for Mixed Discriminants, Advances in Mathematics, 2006.
13. L. Gurvits, van der Waerden/Schrijver-Valiant like Conjectures and Stable (aka Hyperbolic) Homogeneous Polynomials: One Theorem for all, Electron. J. Combin. 15 (2008), no. 1, Research Paper 66, 26 pp.
14. L. Gurvits, Polynomial time algorithms to approximate mixed volumes within a simply exponential factor, arXiv:cs/0702013v3, 2007.
15. L. Gurvits, A Short Proof, Based on Mixed Volumes, of Liggett's Theorem on the Convolution of Ultra-Logconcave Sequences, The Electronic Journal of Combinatorics, Volume 16(1), 2009.
16. L. Hormander, Analysis of Linear Partial Differential Operators, Springer-Verlag, New York, Berlin, 1983.
17. V.L. Kharitonov and J.A. Torres Munoz, Robust Stability of Multivariate Polynomials. Part 1: Small Coefficients Perturbations, Multidimensional Systems and Signal Processing, 10 (1999), 7-20.
18. A.G. Khovanskii, Analogues of the Aleksandrov-Fenchel inequalities for hyperbolic forms, Soviet Math. Dokl. 29 (1984), 710-713.
19. H. Minc, Permanents, Addison-Wesley, Reading, MA, 1978.
20. C. Niculescu, "A New Look at Newton's Inequalities". Journal of Inequalities in Pure and Applied Mathematics 1 (2) (2000).
21. A. Okounkov, Why would multiplicities be log-concave? The orbit method in geometry and physics (Marseille, 2000), 329–347, Progr. Math., 213, Birkhäuser Boston, Boston, MA, 2003.
22. A. Schrijver, Counting 1-factors in regular bipartite graphs, Journal of Combinatorial Theory, Series B 72 (1998) 122–135.
23. G.C. Shephard, Inequalities between mixed volumes of convex sets, Mathematika 7 (1960), 125-138.

Chapter 5
Niceness theorems

Michiel Hazewinkel

Abstract There are many results and constructions in mathematics that are $*$un-reasonably nice$*$. For instance it appears to be difficult for a set to carry many compatible (algebraic) structures. More precisely, if, say, an algebra carries a compatible $*$higher$*$ structure the underlying algebra must be very regular. For instance, if an associative unital algebra (over a characteristic zero field) carries a graded connected Hopf algebra structure the underlying algebra is free commutative. There are many such theorems in various different parts of mathematics. This paper gives a number of examples of this phenomenon and of similar phenomena as a preliminary step in starting to examine and try to understand this matter. Besides unreasonably nice constructions and theorems there is also the matter of nice proofs. By this I mainly mean proofs that principally rely on, for instance, the universal properties that define an object, and that do not rely (too much) on calculations. This matter is touched upon in the last section of this paper.

5.1 Introduction and statement of the problems

In this paper I aim to raise a new kind of question.[1] It appears that many important mathematical objects (including counterexamples) are unreasonably nice, beautiful and elegant. They tend to have (many) more (nice) properties and extra bits of structure than one would a priori expect.

Michiel Hazewinkel
Burg. s'Jacob laan 18 1401BR Bussum The Netherlands
e-mail: michiel.hazewinkel@xs4all.nl

[1] The first time I lectured on this at the WWCA conference, Wilfrid Laurier University in Waterloo, Canada, May 2008, the chairman summed up my lecture as follows: "If it is true it is beautiful, if it is beautiful it is probably true". I also lectured on the same subject at the Abel symposium meeting in Tromsø, Norway in June 2008. The present screed expands on those first lectures a great deal. Yet, in spite of its length it is just a beginning: a fist scratching at the edges of a great and fascinating problem that deserves devoted attention.

I.S. Kotsireas, E.V. Zima (eds.), *Advances in Combinatorial Mathematics*,
DOI 10.1007/978-3-642-03562-3_5,

The question is why this happens and whether this can be understood.[2]

These ruminations started with the observation that it is difficult for, say, an arbitrary algebra to carry additional compatible structure. To do so it must be nice, i.e., as an algebra be regular (not in the technical sense of this word), homogeneous, everywhere the same, ... It is for instance very difficult to construct an object that has addition, multiplication and exponentiation, all compatible in the expected ways.

The present scribblings are just a first attempt to identify and describe the phenomenon. Basically this is a prepreprint and it touches just the fringes of the subject. There is much more to be said and there are many more examples than remarked upon here.

This paper is about lots of examples of this phenomenon such as Daniel Kan's observation that a group carries a comonoid structure in the category of groups if and only if it is a free group, the Milnor-Moore and Leray theorems in the theory of Hopf algebras, Grassmann manifolds and classifying spaces, and especially the star example: the ring of commutative polynomials over the integers in countably infinite indeterminates. This last one occurs all over the place in mathematics and has more compatible structures that can be believed. For instance it occurs as the algebra of symmetric functions in infinitely many variables, as the cohomology and homology of the classifying space **BU**, as the sum of the representation rings of the symmetric groups, as the free lambda-ring on one variable, as the representing ring of the Witt vectors, as the ring of rational representation of GL_∞, as the underlying ring of the universal formal group, ...

To start with, here is a preliminary list of the kind of phenomena I have in mind.

- A. Objects with a great deal of compatible structure tend to have a nice regular underlying structure and/or additional nice properties: "Extra structure simplifies the underlying object". As indicated above this sort of thing was the starting point.

- B. Universal objects. That is mathematical objects which satisfy a universality property. They tend to have:

a) a nice regular underlying structure
b) additional universal properties (sometimes seemingly completely unrelated to the defining universal property)

---

[2] There is of course the "anthropomorphic principle" answer, much like the question of the existence of (intelligent) life in this universe. It goes something like this. If these objects weren't nice and regular we would not be able to understand and describe them; we can see/understand only the elegant and beautiful ones. I do not consider this answer good enough though there is something in it. So the search is also on for ugly brutes of mathematical objects. Also this anthropomorphic argument raises the subsidiary questions of why we can only understand/describe beautiful/regular things. There are aspects of (Kolmogorov) complexity and information theory involved here.

- C. Nice objects tend to be large and inversely large objects of one kind or another tend to have additional nice properties. For instance, large projective modules are free [12]

- D. Extremal objects tend to be nice and regular. (The symmetry of a problem tends to survive in its extremal solutions is one of the aspects of this phenomenon; even when (if properly looked at) there is bifurcation (symmetry breaking) going on.)

- E. Uniqueness theorems and rigidity theorems often yield nice objects (and inversely). They tend to be unreasonably well behaved. I.e. if one asks for an object with such and such properties and the answer is unique the object involved tends to be very regular. This is not unrelated to D.

Concrete examples of all these kinds of phenomena will be given below (section 2) as well as a (pitiful) few first explanatory general theorems (section 3). The "niceness phenomenon" is not limited to theorems saying that e.g. in suitable circumstances an object is free; it also extends to counter examples: many of them are very regular in their construction. This can, for instance, take the form of a simple construction repeated indefinitely. Some examples are in section 5.2.6 below. All in all I detect in present day mathematics a strong tendency towards the study of things that in some sense have low Kolmogorov complexity.

5.2 Examples

5.2.1 Lots of compatible structure examples

5.2.1.1 Groups in the category of groups

To start with here is an observation of Daniel Kan, [89], which has moreover the distinction of being one of the first results of this kind and of admitting a nice (sic!) pictorial illustration.

First, here is the abstract setting. Let $\mathscr{C}$ be a category with a terminal object and products. For example the category **Group** of groups where the product is the direct product and the terminal object is the one element group.

A group object in such a category $\mathscr{C}$ is an object $G \in \mathscr{C}$ equipped with a morphism $m : G \times G \to G$ (multiplication), a morphism $e : T \to G$ (unit element) where T is the terminal object of the category $\mathscr{C}$, and a morphism $\iota : G \to G$ (inverse) such that the categorical versions of the standard group axioms hold. This means that the following diagrams are supposed to be commutative.

$$\begin{array}{ccc} G\times G\times G & \xrightarrow{m\times id} & G\times G \\ \downarrow{\scriptstyle id\times m} & & \downarrow{\scriptstyle m} \\ G\times G & \xrightarrow{m} & G \end{array} \quad \text{(associativity)} \tag{5.1}$$

$$\begin{array}{ccc} G\times T & \xrightarrow{id\times e} & G\times G \\ \uparrow{\scriptstyle \cong} & & \downarrow{\scriptstyle m} \\ G & \xrightarrow{=} & G \end{array} \qquad \begin{array}{ccc} T\times G & \xrightarrow{e\times id} & G\times G \\ \uparrow{\scriptstyle \cong} & & \downarrow{\scriptstyle m} \\ G & \xrightarrow{=} & G \end{array} \quad \text{(unit)} \tag{5.2}$$

$$\begin{array}{ccc} G & \xrightarrow{(id,\iota)} & G\times G \\ \downarrow & & \downarrow{\scriptstyle m} \\ T & \xrightarrow{e} & G \end{array} \qquad \begin{array}{ccc} G & \xrightarrow{(\iota,id)} & G\times G \\ \downarrow & & \downarrow{\scriptstyle m} \\ T & \xrightarrow{e} & G \end{array} \quad \text{(inverse)} \tag{5.3}$$

where the vertical arrow on the left hand side of the two diagrams (5.3) is the unique morphism in the category $\mathscr{C}$ to the terminal object and the vertical isomorphisms on the left of (5.2) are the canonical isomorphisms of an object with the product of that object with the terminal object.

In the case of the category of groups this means that a group object is a group with composition law denoted $+$ (though it is not clear yet that it is commutative) with a second composition law, denoted $*$ that is distributive over the first composition law in the sense that the following identity holds

$$(a+b)*(a'+b') = (a*a')+(b*b') \tag{5.4}$$

This comes from the requirement that $*$ must be a morphism in the category **Group**. Let 0 be the unit element for the composition law $+$ and 1 the unit element for the composition law $*$. Putting $b = a' = 0$ in (5.4) gives

$$a*b' = (a*0)+(0*b') \tag{5.5}$$

On the other hand putting $a' = b' = 1$ in (5.4) gives

$$a+b = (a+b)*(1+1)$$

and multiplying this with the inverse of $(a+b)$ for the star composition gives $1 = 1+1$ and hence $1 = 0$. Put this in (5.5) to find that $a*b' = a+b'$ showing that the compositions are the same and then (5.4) immediately gives that both are Abelian.

Thus a group object in the category of groups is Abelian and the second composition law is the same as the first. Actually this can be proved more generally for monoid objects in the category of groups [89].

There is a nice illustration of this in homotopy theory (and that is where the idea came from). This goes as follows. The second homotopy group, $\pi_2(X,*)$, of a based space $(X,*)$ is, as a set, the set of all homotopy classes of maps from the disk into X that take the boundary circle into the base point $*$ of X.

For illustrational (and conceptual) purposes it is easier to think of homotopy classes of maps from the unit filled square to X that take the boundary to the base point. Homotopically, of course, this makes no difference.

Now let

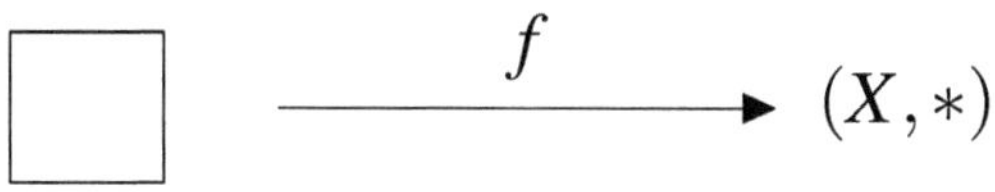

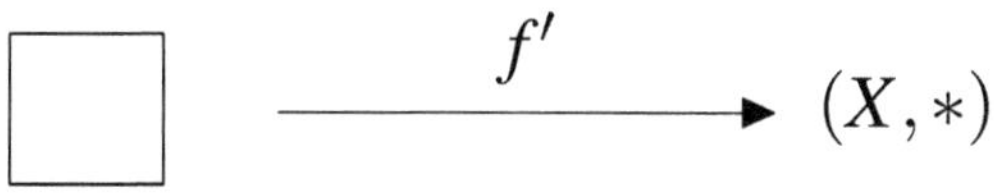

be two such maps. They can be glued together horizontally to give a map of the same kind (up to homotopy):

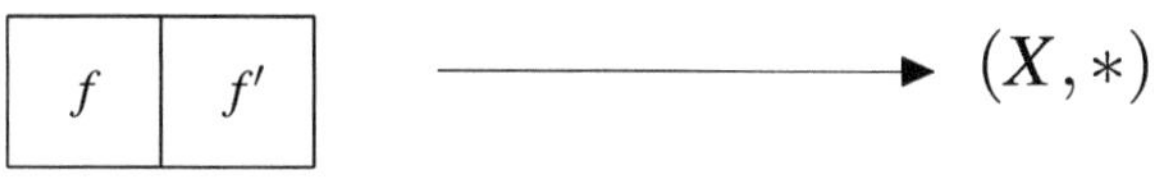

and this induces a composition on $\pi_2(X,*)$ turning it into a group. Of course the two maps can also be glued together vertically, inducing another, a priori different, group structure.

Now take four such maps f, f', g, g'. Then first gluing f, f' and g, g' together horizontally and then gluing the two results together vertically gives a map that can be depicted

Obviously the same result is obtained by first gluing f and g together vertically, gluing f' and g' together vertically, and gluing the results together horizontally. This establishes the relation (5.4) in the present case and shows that $\pi_2(X,*)$ is Abelian.[3]

[3] In the present case of homotopy groups it can of course also easily be shown directly that vertical gluing and horizontal gluing give the same result and this is how things are done traditionally in textbooks; see e.g. [84].

5.2.1.2 Comonoids in the category of groups

Dually there is the notion of a cogroup object in a category.

For this let $\mathscr{C}$ be a category with direct sums and an initial object. Again the category of groups is an example with the one element group as initial object. The categorical direct sum in **Group** is what in group theory is called the free product.

A cogroup object in such a category $\mathscr{C}$ is an object $C \in \mathscr{C}$ together with a comultiplication $\mu : C \to C\Sigma C$, a coinverse $\iota : C \to C$, and a counit morphism $\varepsilon : C \to I$. Here $I \in \mathscr{C}$ is the initial object and Σ stands for the direct sum in $\mathscr{C}$. These bits of structure are supposed to satisfy the dual axioms to those for a group object depicted by diagrams (5.1) – (5.3), that is the diagrams obtained by reversing all arrows (and replacing m by μ and e by ε) must be commutative. For a comonoid object leave out the coinverse and (the dual of) diagram (5.3).

It is now a theorem, [89], that the underlying group of a comonoid object in the category of groups is free as a group. This has much to do with the fact that the categorical direct sum in **Group** is given by the free product construction.

5.2.1.3 Hopf's theorem on the cohomology of H-spaces

An H-space is a based topological space $(X,*)$ together with a continuous map $m : X \times X \to X$ such that $x \mapsto m(x,*)$ and $x \mapsto m(*,x)$ are homotopic to the identity.[4] The result of Heinz Hopf, [82], see also [50], alluded to is now as follows.

Let k be a field of characteristic zero and X a path connected H-space such that $H_*(X;k)$ is of finite type then $H^*(X;k)$ is a free graded-commutative graded algebra.

Here 'finite type' means that each $H_i(X;k)$ is finite dimensional and the cohomology algebra is graded-commutative (= commutative in the graded sense), i.e. $xy = (-1)^{degree(x)degree(y)}yx$. Thus the seemingly weak extra bit of structure 'H-space' has a profound influence on the (cohomological) structure of a space.

5.2.1.4 Intermezzo: Hopf algebras

Let R be a unital commutative ring. A graded module over R is simply a collection of modules over R indexed by the nonnegative integers.[5] Or, equivalently, it is a direct sum

$$M = \bigoplus_{i \in N \cup \{0\}} M_i \tag{5.6}$$

An element $x \in M_i$ is said to be homogeneous of degree i. A graded module (5.6) is said to be of finite type if each of the M_i is of finite rank over the base ring R.

---

[4] Often in the literature for an H-space it is also required that the 'multiplication' m is associative up to homotopy. For the present result that is not required.

[5] These will be the only kind of gradings occurring.

The tensor product of two graded modules M,N is graded by assigning degree $i+j$ to the elements from $M_i \otimes N_j$.

A graded algebra over R is a graded module (5.6) equipped with a graded associative multiplication and a unit element

$$m : M \otimes M \to M,\ m(M_i \otimes M_j) \subset M_{i+j};\ 1 \in M_0 \tag{5.7}$$

There are two notions of commutativity for graded algebras: (ordinary) commutativity, which means $xy = yx$, and graded-commutativity, which means $xy = (-1)^{\deg(x)\deg(y)} yx$. Both occur frequently in the literature and both will occur in the present paper.[6]

Correspondingly there are two versions for the multiplication in the tensor product of (the underlying graded modules) of graded rings, viz.

$$\begin{aligned} (x \otimes y)(x' \otimes y') &= xx' \otimes yy' \\ (x \otimes y)(x' \otimes y') &= (-1)^{\deg(y)\deg(x')} xx' \otimes yy' \end{aligned} \tag{5.8}$$

where in the second equation the elements $x, x'y, y'$ are supposed to be homogeneous. The sign factor in the second equation of (5.8) is needed to ensure that the tensor product of two graded-commutative graded algebras is graded-commutative (as of course one wants it to be).

Dually a graded coalgebra over R is a graded module equipped with a coassociative comultiplication and a counit

$$\mu : M \to M \otimes M, \mu(M_n) \subset \bigoplus_{i+j=n} M_i \otimes M_j; \varepsilon : M \to R, \varepsilon(M_i) = 0 \text{ for } i > 0$$

Just as in the algebra case there are two notions of cocommutativity and two ways to define a coalgebra structure on the tensor product of two graded coalgebras. These two are as follows. Let C and D be two graded coalgebras with comultiplications μ_C, μ_D. Write

$$\mu_C(x) = \sum x'_i \otimes x''_i,\ \mu_D(y) = \sum y'_j \otimes y''_j$$

as sums of tensor products of homogeneous elements. Then the two graded coalgebra structures alluded to are

$$\begin{aligned} x \otimes y &\mapsto \Sigma x'_i \otimes y'_j \otimes x''_i \otimes y''_j \\ x \otimes y &\mapsto \Sigma (-1)^{\deg(y'_j)\deg(x''_i)} x'_i \otimes y'_j \otimes x''_i \otimes y''_j \end{aligned} \tag{5.9}$$

Next, a graded bialgebra B is a comonoid object in the category of graded algebras or, equivalently, a monoid object in the category of graded coalgebras. Here again there are two versions depending on what algebra and coalgebra structures are taken on $B \otimes B$. First there is an ‘ordinary’ bialgebra which happens to carry a grad-

[6] If all the odd degree summands of the graded ring are zero the two notions agree. This can be used to unify things.

ing. In this case the algebra and coalgebra structures are given by the first formulas of (5.8) and (5.9). Second there is the 'grade-twist' version in which the algebra and coalgebra structures on the tensor product are given by the second formulas from (5.8) and (5.9). Here 'ordinary twist' and 'grade twist' respectively refer to the morphisms

$$x \otimes y \mapsto y \otimes x, \; x \otimes y \mapsto (-1)^{\deg(x)\deg(y)} y \otimes x$$

which make their appearance when the conditions are written out explicitly in terms of diagrams or elements.

Finally, a graded Hopf algebra is a graded bialgebra that in addition carries a so-called antipode. That is a morphism ι of graded modules of degree 0 (so that $\iota(H_i) \subset H_i$) that satisfies

$$m(id \otimes \iota)\mu = e\varepsilon \quad \text{and} \quad m(\iota \otimes id)\mu = e\varepsilon.$$

A graded Hopf algebra over R is connected if the grade zero part H_0 is equal to R so that e and ε induce isomorphism of R with H_0.

An element x in a graded Hopf algebra (or bialgebra) is called primitive if it satisfies

$$\mu(x) = 1 \otimes x + x \otimes 1 \tag{5.10}$$

These form a graded submodule $P(H)$ of the Hopf algebra H. In the case of an 'ordinary twist' Hopf algebra the commutator product

$$[x,y] = xy - yx \tag{5.11}$$

turns $P(H)$ into a Lie algebra (that happens to carry a grading such that the Lie bracket is of degree 0). In the case of a 'graded twist' Hopf algebra take

$$[x,y] = xy - (-1)^{\deg(x)\deg(y)} yx \tag{5.12}$$

to obtain a graded Lie algebra. That is a module equipped with a bilinear product [,] that satisfies graded anticommutativity and the graded Jacobi identity:

$$\begin{aligned} &[x,y] = (-1)^{\deg(x)\deg(y)}[y,x] \\ &[x,[y,z]] = [[x,y],z] + (-1)^{\deg(x)\deg(y)}[y,[x,z]] \end{aligned} \tag{5.13}$$

5.2.1.5 Milnor-Moore theorem (topological incarnation)

Let PX be the Moore path space of a path connected based topological space $(X,*)$. That is the space of paths starting from $*$ with specified length (which is what the adjective 'Moore' means in this context). Assigning to a path its endpoint defines a continuous map $PX \to X$, which is a fibration with ΩX, the space of Moore loops, as its fibre (over $*$). As PX is contractible the long exact homotopy sequence attached to this fibration gives isomorphisms $\pi_n(X) \to \pi_{n-1}(\Omega X)$. This can be used to transfer the Whitehead products $\pi_m(X) \times \pi_n(X) \to \pi_{m+n=1}(X)$ to a Lie product (of degree zero) $(\pi_*(\Omega X)\otimes k)\times(\pi_*(\Omega X)\otimes k)\xrightarrow{m_{\Omega X}}\pi_*(\Omega X)\otimes k$, defining a graded Lie algebra L_X.

Composition of loops turns ΩX into a topological monoid and, up to homotopy there is an inverse as well. Using the Alexander - Whitney and Eilenberg - Zilber chain complex equivalences, see [50], p. 53ff, and the fact that taking homology of chain complexes commutes with tensor products, ibid. p. 48, the composition $\Omega X\times\Omega X\to\Omega X$ and diagonal $\Delta:\Omega X\to\Omega X\times\Omega X$ induce an algebra and coalgebra structure on $H_*(\Omega X)$. Moreover, essentially because a loop in a product $X\times Y$ is a pair of loops and composition of loops seen this way goes component-wise, the comultiplication morphism $H_*(\Omega X)\to H_*(\Omega X)\otimes H_*(\Omega X)$ is an algebra morphism,[7] ibid. p. 225.

All in all this turns $H_*(\Omega X)$ into a graded connected Hopf algebra (of the 'graded twist' kind).

Now let the coefficients ring used when taking cohomology be a field of characteristic zero.

5.2.1.6 Theorem ([127], see also [50], p. 293)

Let X be a simply connected path connected topological space. Then the Hurewicz homomorphism for ΩX is an isomorphism of graded Lie algebras of L_X onto the graded Lie algebra of primitives of $H_*(\Omega X;k)$ and this isomorphism extends to an isomorphism of graded Hopf algebras of the universal enveloping algebra UL_X with $H_*(\Omega X;k)$.[8]

There is also a purely algebraic theorem that goes by the name 'Milnor-Moore theorem'. That one involves the notion of the universal enveloping algebra of a Lie algebra and will be discussed in subsection 5.2.3 below. To conclude this section 5.2.1 let me briefly mention two more simple results that, I feel, qualify as 'niceness theorems'. Both say that the presence of a Hopf algebra (bialgebra) structure has implications for the underlying algebra.

5.2.1.7 Cartier's theorem on nilpotents in group schemes

Let H be a finite dimensional Hopf algebra over a field of characteristic zero. Then the underlying algebra has no nilpotents. Actually a much stronger statement holds, see [44]. The usual statement is: A group scheme of finite type over a field of characteristic zero is smooth. See loc. cit. and [159], p. 7.

In characteristic $p>0$, Cartier's theorem does not hold. On $k[X]/(X^p)$ where k is a field of characteristic $p>0$, there are the two comultiplications

---

[7] This is the origin of the unfortunate but frequently used notation 'Δ' for the comultiplication in a Hopf algebra.

[8] Universal enveloping algebras are the topic of section 5.2.2.1 below.

$$X \mapsto 1 \otimes X + X \otimes 1, \quad X \mapsto 1 \otimes X + X \otimes 1 + X \otimes X$$

and both define a bialgebra, and in fact Hopf algebra structure on $k[X]/(X^p)$. These two Hopf algebras (finite group schemes) are traditionally denoted α_p and μ_p.

5.2.1.8

Let k be a field and n an integer ≥ 2. Then there is no bialgebra structure on the algebra $M^{n\times n}(k)$ of $n \times n$ matrices over k. See [40], p. 173.

It is completely unknown which products of matrix algebras do carry (admit) a bialgebra structure.

Much of mathematics concerns statements as to what consequences follow from what assumptions. So it can be argued that there is nothing particularly special about the results described above. However, I feel that there is something special, something particularly elegant, about the results described. Part of the general problem is to understand why and in what sense.

Several of the theorems above are 'freeness theorems'. They say that in the presence of suitable extra structure an object is free. Here follow five more. For the first three the 'extra structure' is that the object in question is imbedded in a free object. In some categories that means nothing; in others it is a strong bit of extra structure. Just what categorical properties rule this behavior is completely unknown.

5.2.1.9 Nielsen-Schreier theorem

A subgroup of a free group is free, [142] [133]; [150], p. 181.

5.2.1.10 Shirshov-Witt theorem

Lie subalgebras of a free Lie algebra are free, [147], [162]. There is also, up to a point, a braided version, [93].

5.2.1.11 Bergman centralizer theorem

The centralizer of a non-scalar element in a free power series ring $k \ll X \gg$ is of the form $k[[c]]$, [17], [38], p. 244.. Here c is a single element!

5.2.1.12

The fundamental group of a cogroup object in the homotopy category of 'nice' based topological spaces is free. See [18]. These objects are sometimes called H'-spaces (as a kind of dual or opposite object to H-spaces).

5.2.1.13 Bott-Samelson theorem

The homology algebra $H_*(\Omega\Sigma X;k)$ is a free algebra generated by $H_*(X;k)$, [24], [18]. Here Σ is the suspension functor and Ω is the loop space functor on based topological spaces. These are adjoint and there results a topological morphism $X \to \Omega\Sigma X$. The multiplication comes from the fact that loops at the base point can be composed making a loop space an H-space.

5.2.2 Universal object examples

Here the theme is that objects that are defined in terms of some universal property have a tendency to pick up extra bits of structure.

5.2.2.1 The universal enveloping algebra of a Lie algebra

Let A be a unital associative algebra over a unital commutative base algebra R. Associated to A there is a Lie algebra structure on A defined by the commutator difference

$$[x,y]_A = xy - yx \tag{5.14}$$

Let $\mathfrak{g}$ be a Lie algebra. A Lie morphism from $\mathfrak{g}$ to a unital associative algebra A is a module morphism $\phi : \mathfrak{g} \to A$ such that $\phi([x,y]_{\mathfrak{g}}) = [\phi x, \phi y]_A$. The universal enveloping algebra on $\mathfrak{g}$ is a unital associative algebra $U\mathfrak{g}$ together with a Lie morphism $i : \mathfrak{g} \to U\mathfrak{g}$ such that for each Lie morphism $\phi : \mathfrak{g} \to A$ there is a unique morphism of associative algebras $\tilde{\phi} : U\mathfrak{g} \to A$ such that $\tilde{\phi} \circ i = \phi$. Pictorially (in diagram form) this can be rendered as follows

$$\begin{array}{ccc} \mathfrak{g} & \longrightarrow & U\mathfrak{g} \\ & {\scriptstyle \phi}\searrow \quad \swarrow{\scriptstyle \exists^1 \tilde{\phi}} & \\ & A & \end{array} \tag{5.15}$$

The associative unital algebra $U\mathfrak{g}$ is a very nice one. For instance there is the Poincaré - Birkhoff - Witt theorem that specifies (under suitable circumstances) a monomial basis for it. This results basically from the construction of $U\mathfrak{g}$. (And one wonders whether this PBW theorem can be deduced directly from the characterizing universality property.)

What is of interest in the present setting is that the universality property immediately implies that $U\mathfrak{g}$ has more structure; in fact that it is a Hopf algebra. This arises as follows. Consider the associative algebra $U\mathfrak{g} \otimes U\mathfrak{g}$ and the morphism $x \mapsto 1 \otimes x + x \otimes 1$ from $\mathfrak{g}$ into it. It is immediate that this is a Lie morphism and hence there is a corresponding (unique) morphism of associative algebras $U\mathfrak{g} \to U\mathfrak{g} \otimes U\mathfrak{g}$. It is immediate that this turns $U\mathfrak{g}$ into a Hopf algebra.

There is a completely analogous picture for graded Lie algebras.

Of course the universal problem described here is an instance of an adjoint functor situation. Let **Lie** be the category of Lie algebras (over R) and **Alg** the category of unital associative algebras (over R). Then associating to an associative algebra A its commutator difference product is a (forgetful) functor $V : \mathbf{Alg} \to \mathbf{Lie}$ and $\mathfrak{g} \to U\mathfrak{g}$ is a functor the other way that is left adjoint to it:

$$\mathbf{Lie}(\mathfrak{g}, V(A)) \cong \mathbf{Alg}(U\mathfrak{g}, A) \tag{5.16}$$

In the case of a forgetful functor a left adjoint to it yields what are often called free objects (as in this case). Thus $U\mathfrak{g}$ is the free associative algebra on the Lie algebra $\mathfrak{g}$.

A right adjoint functor to a forgetful functor gives cofree objects. An example of a cofree construction will occur below.

The very important notion of adjointness is due to Daniel Kan, [88] and as Saunders Mac Lane says in the preface of [116] "Adjoint functors arise everywhere".

If (F,G) is an adjoint functor pair, i.e. e.g. $\mathfrak{C}(FX,Y) \cong \mathfrak{D}(X,GY)$ functorialy (loosely formulated), one expects niceness properties for both the FX's and the GY's. And indeed many niceness results fall into this scope with the proviso that often these objects pick up extra properties which are not implicit in the adjoint situation alone.

5.2.2.2 The group algebra of a group

Much the same picture holds for the group algebra of a group. Except much easier. Here the 'forgetful functor' assigns to an algebra A its group A^* of invertible elements. Recall that the group algebra kG of a group is the free module over k with basis G and the multiplication determined on this basis by the group multiplication. The adjointness equation now is:

$$\mathbf{Group}(G, A^*) \cong \mathbf{Alg}_k(kG, A) \tag{5.17}$$

There is again a Hopf algebra structure for free. For this, to put things formally on the same footing as in the case of the universal enveloping algebra, consider the morphism

$$G \to (kG \otimes kG)^*, \quad g \mapsto g \otimes g$$

which by the adjointness equation (5.17), gives rise to a morphism of algebras $kG \to kG \otimes kG$ turning kG into a bialgebra (and a Hopf algebra using the group inverse). Of course in this case things are so simple that it is not worthwhile to go through this yoga.

5.2.2.3 Free algebras

Everyone knows how to construct the free algebra over a module (or a set). The tensor algebra does the job and that is a very nice structure. Less known is that this also works in the setting **CoAlg - HopfAlg**, where **CoAlg** and **HopfAlg** are suitable categories of coalgebras and Hopf algebras over a suitable base ring. See [130] and [70]. This gives the free Hopf algebra on a coalgebra.

5.2.2.4 Cofree coalgebras

Given a module M, the cofree coalgebra over[9] M would be a coalgebra $C(M)$ together with a module morphism $C(M) \xrightarrow{\eta} M$ such that for each coalgebra C together with a morphism of modules $C \xrightarrow{\alpha} M$ there is a unique morphism of coalgebras $\widehat{\alpha} : C \to C(M)$ such that $\eta\widehat{\alpha} = \alpha$.

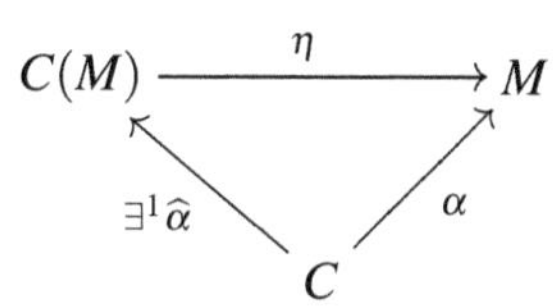

Whether the cofree coalgebra over a module always exist is not quite settled, [76]; they certainly exist in many cases. In the connected graded context they always exist and are given by the tensor coalgebra, again a very nice structure.

And in this connected graded context there is the **Alg - HopfAlg** version giving the cofree Hopf algebra over an algebra, [70], [130].

5.2.2.5 The classifying spaces $\mathbf{BU}_n$

A completely different kind of universal object is formed by the complex Grassmannians and their inductive limits the classifying spaces $\mathbf{BU}_n$.

Consider the complex vector space $\mathbf{C}^{n+r}$ and define the complex Grassmannian

$$\mathbf{Gr}_n(\mathbf{C}^{n+r}) = \{V : V \text{ is an } n-\text{dimensional subspace of } \mathbf{C}^{n+r}\} \tag{5.18}$$

This set has a natural structure of a smooth manifold (in fact a complex analytic manifold). Letting r go to infinity (which technically means taking an inductive limit) gives the classifying space

$$\mathbf{BU}_n = \varinjlim \mathbf{Gr}_n(\mathbf{C}^{n+r}) = \mathbf{Gr}_n(\mathbf{C}^\infty) \tag{5.19}$$

---

[9] It pays to be terminologically careful in this context. I prefer to speak of the free algebra *on* a module and the cofree coalgebra *over* a module.

It is also perfectly possible to define and work directly with the most right hand side of (5.19). There is a canonical complex vector bundle over $\mathbf{BU}_n$ which is colloquially defined by saying the fibre over $x \in \mathbf{BU}_n$ "is x". More precisely this canonical vector bundle γ_n is

$$\gamma_n = \{(x,v) : x \in \mathbf{BU}_n, v \in x\} \text{ with projection } (x,v) \mapsto x, \gamma_n \to \mathbf{BU}_n$$

There is now the following universality/classifying property. For every paracompact space X with an n-dimensional complex vector bundle ξ over it there is a map $f_\xi : X \to \mathbf{BU}_n$ such that ξ is isomorphic (as a vector bundle) to the pullback $f_\xi^*(\gamma_n)$. Moreover f_ξ is unique up to homotopy.

The remarkable thing here is that the classifying spaces $\mathbf{BU}_n$ are so elegant and simple (as are the universal bundles over them). There are more nice properties. Jumping the gun a little – these spaces will return later – the cohomology of these spaces is particularly nice

$$H^*(\mathbf{BU}_n;\mathbf{Z}) = \mathbf{Z}[c_1, c_2, \ldots, c_n], \; \deg(c_r) = 2r \tag{5.20}$$

All this can be found in [86], [128] (and many other books).

5.2.3 Niceness theorems for Hopf algebras

The structure of a Hopf algebra is a heavy one. Indeed at one time they were thought to be so rare that each and every one deserves the most careful study, [90]. This is not the case anymore. Hopf algebras abound. Still the structure is not strong enough to produce good niceness theorems. However if one adds conditions like graded and connected some strong structure theorems emerge. These are e.g. the Leray and Milnor – Moore theorems which will both be described immediately below. In addition there is the Zelevinsky theorem, a structure theorem due to Grünenfelder, [66] and much more, see e.g. [123]. However, whether the various available classification theorems for Hopf algebras qualify as niceness theorems is debatable. I think mostly not.

5.2.3.1 The Leray theorem on commutative Hopf algebras

Let H be a commutative graded connected Hopf algebra of finite type over a field of characteristic zero. Then the underlying algebra is commutative free. There is also a graded commutative version. The original theorem appears in [109]. For an up-to-date short account see [136]. There are all kinds of generalizations, e.g. to an operadic setting, see [135], [113], [57].

5.2.3.2 The Milnor – Moore theorem on cocommutative Hopf algebras

Let H be a cocommutative graded connected Hopf algebra of finite type over a field of characteristic zero. Then the underlying algebra is the universal enveloping algebra of the Lie algebra of primitives $P(H)$ of H, [127].

This is the algebraic incarnation referred to in 5.2.1.6 above.

The Milnor–Moore theorem is a dual of the Leray theorem. To realize this recall from subsection 5.2.2 above that $U\mathfrak{g}$ is the free object in **Ass** on the object $\mathfrak{g} \in$ **Lie**.

5.2.4 Large vs nice

There is a tendency for (really) nice objects to be big (or very small). A prime example is

$$\mathbf{Symm} = \mathbf{Z}[h_1, h_2, h_3, \ldots] \tag{5.21}$$

the ring of polynomials over the integers in countably infinite many commuting variables over the integers. This object will be discussed in some detail further on.

Inversely big objects have a better chance of being nice.

In this subsection I give some examples of this phenomenon.

5.2.4.1 Big projective modules are free

This result is due to Hyman Bass, [12]. For a precise statement see loc. cit. (corollary 3.2). The key ingredient is the following elegant observation.[10]

If $P \oplus Q \cong F$ with F a non-finitely generated free module, then $P \oplus F \cong F$.

The proof is simplicity itself and clearly shows the power and usefulness of infinity.

$$\begin{aligned} F &\cong F \oplus F \oplus \cdots \cong P \oplus Q \oplus P \oplus Q \oplus \cdots \\ &\cong P \oplus F \oplus F \oplus \cdots \cong P \oplus F \end{aligned}$$

5.2.4.2 General linear groups in various dimensions

Let k be the field of real numbers, complex numbers or even the quaternions. The general linear groups $\mathbf{GL}_n(k)$ for finite natural numbers are homotopically and cohomologically far from trivial.

Things change drastically in infinite dimension.

[10] Hyman Bass calls it “an elegant little swindle”.

5.2.4.3 Kuiper's theorem, [100]

Let H be real or complex or quaternionic Hilbert space. Then the general linear group $\mathbf{GL}(H)$ is contractible.

There is also an important equivariant extension due to Graeme Segal, [146].

Much related is Bessaga's theorem, [19], [20], to the effect that every infinite dimensional Hilbert space is diffeomorphic with its unit sphere.

Kuiper's famous theorem is the key to the classification of Hilbert manifolds, [28], [46], [47], [131], [132].

5.2.4.4

Here is a table on differential topology in various dimensions as things seem to be constituted at present.

1	2	3	4	5	6	...	$< \infty$	∞
Real easy	Easy	difficult	difficult; boapw	good techniques	good techniques	...	good techniques	real nice

Here 'good techniques' refers mainly to Smale's handlebody theory. The acronym 'boapw' means 'best of all possible worlds' and refers to the fact that all $\mathbf{R}^n$ for $n \neq 4$ have a unique differentiable structure, but $\mathbf{R}^4$ has over countably infinite different differentiable structures.[11]

5.2.5 *Extremal objects and niceness*

In the world of optimization theory and variational calculus and analysis it is relatively well known that extremal objects tend to be nice (have lots of symmetry), even when bifurcation occurs.

There are also various notions of minimality in algebra and topology and these also tend to be 'nice'. For instance the Sullivan minimal models for rational homotopy, see [50], are definitely nice.

In the world of operads and PROP's etc. there are by way of example the following theorems, see [126].

- The minimal resolution of $\mathfrak{Ass}$ is a differential graded *free* operad.
- The minimal resolution of $\mathfrak{LieB}$ is a *free* differential graded PROP.

Sullivan minimal models and operads, PROP's etc. are highly technical notions and giving details would take me far beyond the scope and intentions of this paper.

I have no doubt that there are more niceness results for minimal resolutions.[12]

---

[11] This is a fact that tends to make 'multiple world' enthusiasts happy.

[12] There are at least three meanings for the word 'resolution' and the phrase 'minimal resolution' in mathematics: resolution of singularities, resolution of a module in homological algebra, resolution in (automatic) theorem proving. Outside mathematics there are many more additional meanings.

5.2.6 Uniqueness and rigidity and niceness

For instance **Symm**, see (5.21) above and below, is unique and rigid as a coring object in the category of unital commutative rings and **MPR**, the Reutenauer-Malvenuto-Poirier Hopf algebra of permutations is rigid and likely unique, see [77], [78]. And indeed they are very nice objects.

5.2.7 Counterexamples and paradoxical objects

Not only objects and constructions can exhibit the 'niceness phenomenon' but also counterexamples. This subsection contains a few examples of that.

5.2.7.1 The Alexander horned sphere

First the construction as illustrated by the picture below. Take a hollow cylinder closed at both ends and bend around so that the two ends face each other. Now from each end extrude a horn and interlock them as shown; there result two locations of disks facing each other. Repeat ad infinitum.

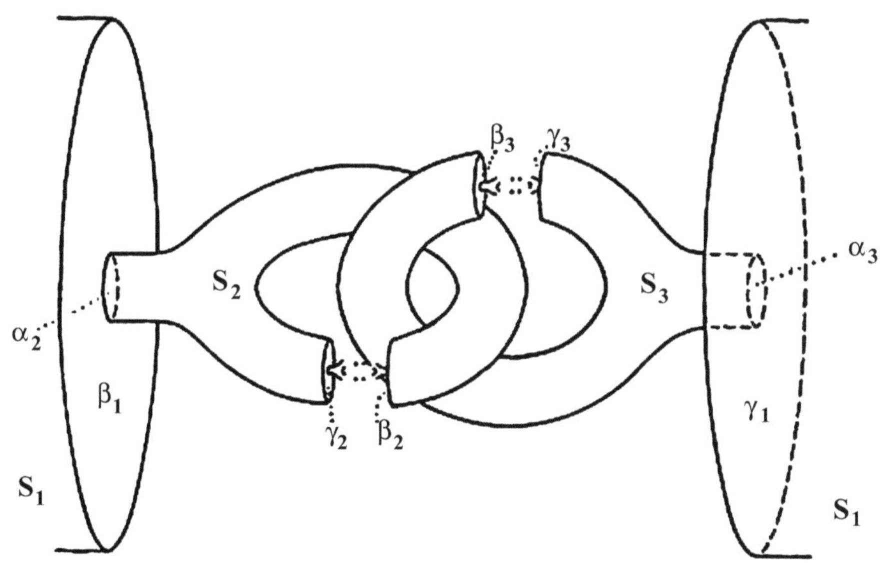

Fig. 5.1 Alexander horned sphere [Credit: MathWorld]

The Alexander horned sphere together with its interior is (homeomorphic to) a topological 3-ball. The exterior is not simply connected. This shows that the analogue of the Jordan-Schönflies theorem from dimension 2 does not hold in dimension 3. For some more information on the Alexander horned sphere and its uses see [1].

Somewhat surprisingly (to me in any case), the filled Alexander horned sphere can be used for a monohedral tiling of $\mathbf{R}^3$, [154].

5.2.7.2 The approximation property

A Banach space is said to have the approximation property if every compact operator is a limit of finite rank operators.

Equivalently a Banach space X has the approximation property if for every compact subset $K \subset X$ and every $\varepsilon > 0$ there is an operator $T : X \to X$ of finite rank such that $\| Tx - x \| < \varepsilon$ for all $x \in K$.

Every Banach space with a (Schauder) basis has the approximation property. This includes Hilbert spaces and the ℓ^p spaces.

However, not every Banach space has the approximation property. In 1973 Per Enflo, [48], constructed a counterexample.

I do not think this counterexample qualifies as a nice one. However the very nice Banach space of bounded operators on ℓ_2 is also a counterexample, [151].[13]

5.2.7.3 The Banach-Tarski paradox

In 1924 Stefan Banach and Alfred Tarski proved the following seemingly bizarre statement, [10].

For two bounded subsets A, B of a Euclidean space of dimension at least three with nonempty interior there exist finite decompositions into disjoint subsets

$$A = A_1 \cup \cdots \cup A_k \quad B = B_1 \cup \cdots \cup B_k$$

such that A_i is congruent to B_i for all $i = 1, \ldots, k$, i.e. A_i becomes B_i under a Euclidean motion.

This is now known as the strong form of the Banach-Tarski paradox. It does not hold in dimensions 1 and 2. A consequence is:

A solid ball can be decomposed into a finite number of point sets that can be reassembled to form two balls identical to the original; see Fig. 5.2 below.

Here 'move' means a Euclidean space move: a combination of translations, rotations and reflections. For some more information on the Banach-Tarski paradox see [160].

Thus 'move' is simple enough. The decomposition, however, is complicated. For one thing at least some of the components must be nonmeasurable. Also things are in three dimensions and Cantor-like sets in three dimensions are difficult to visualize. Fortunately Stan Wagon found a two dimensional analogue in hyperbolic space and the picture is remarkably beautiful; see Fig. 5.3 below.

---

[13] Sometimes 'Szankowski' is rendered 'Shankovskii' which makes it quite hard to find the paper.

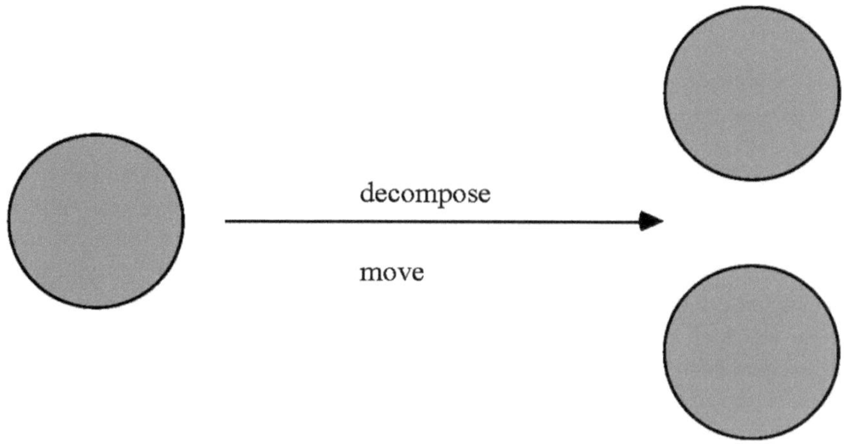

Fig. 5.2 Banach-Tarski paradox: two balls out of one

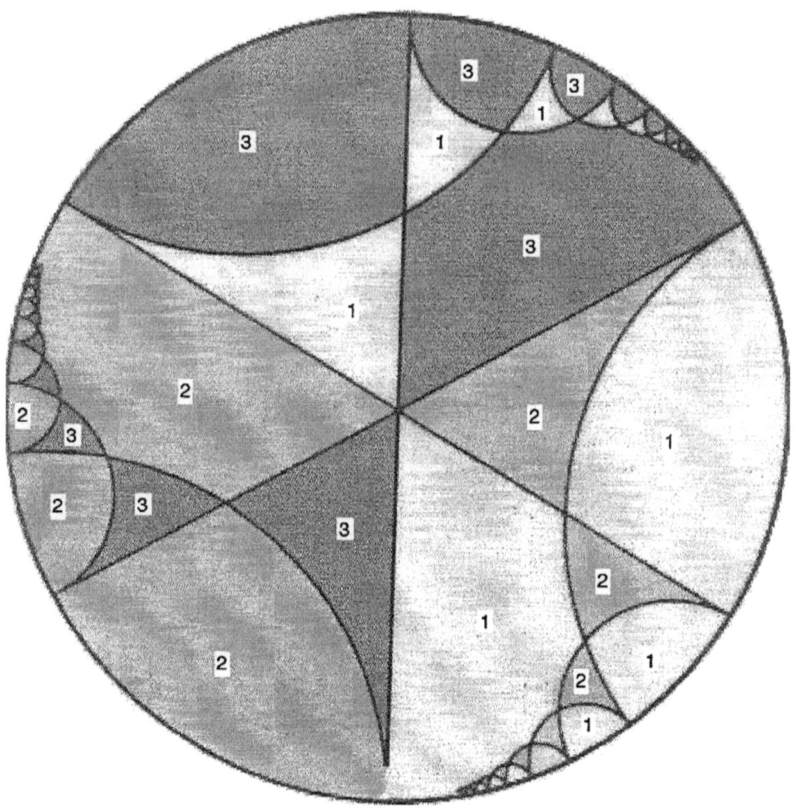

Fig. 5.3 Banach-Tarski paradox: hyperbolic version [Credit: MathWorld]

5.2.7.4 Julia and Fatou sets

Here is a question not untypical of those that were asked in general (point-set) topology almost a century ago when people started to realize just how strange topological spaces could be.

Is it possible to divide the square into three regions so that the boundary between two of them is also the boundary between the two other pairs of regions.

The first answer was given by L. E. J. Brouwer in the form of a simple construction repeated ad infinitum. However, the resulting picture is absolutely not beautiful. Nowadays there are the basins of attraction of discrete dynamical systems such as $x \mapsto x^3 - 1$ which has three basins of attraction (Fatou sets), one for each of the roots of $x^3 - 1$ and each pair has the same boundary (Julia set), see [15].

This is part of the world of fractals and (deterministic) chaos, [145], and many of the pictures are extraordinarily beautiful, [11].[14]

5.2.7.5 Sorgenfrey line

As a set the Sorgenfrey is the set of real numbers. It is given a topology by taking as a basis the half-open intervals $[a,b), a < b$. This topology is finer than the usual one. For instance the sequence $\{n^{-1}\}_{n \in N}$ converges to zero but $\{-n^{-1}\}_{n \in N}$ does not. The Sorgenfrey line serves as a counterexample to several topological properties, [149]. The point here (as far as this paper is concerned) is not that such counterexamples exist but that there is such a nice regular one. There is also a Sorgenfrey plane, loc. cit. For some more information see also [115].

5.2.7.6 Exotic spheres

A further example that fits in this section is that of exotic spheres (Milnor spheres). This deals with existence of differentiable structures on topological spheres, especially the seven dimensional ones, that differ from the standard one. They were the first examples of this phenomenon of distinct differentiable structures on the same topological manifold. This topic is rather more technical, and so I content myself with giving a reference to an internet accessible document, [141].

5.2.8 An excursion into formal group theory

A one dimensional formal group law over a commutative unital ring A is a power series $F(X,Y)$ in two variables with coefficients in A such that

[14] Beautiful and arresting enough that the Sparkasse in Bremen organized an exhibition of them in 1984.

$$F(X,0)=X,\, F(0,Y)=Y,\, F(X,F(Y,Z))=F(F(X,Y),Z) \tag{5.22}$$

Two examples are the multiplicative formal group law and the additive formal group law

$$\widehat{\mathbf{G}}_m(X,Y)=X+Y+XY, \quad \widehat{\mathbf{G}}_\alpha(X,Y)=X+Y \tag{5.23}$$

Both examples are nontypical in that they are polynomial; polynomial formal group laws are very rare.

More generally for any n, including $n=\infty$, an n-dimensional formal group over A is an n-tuple of power series in two groups of n indeterminates $F(X;Y)$ such that

$$F(X;0)=X,\, F(0;Y)=Y,\, F(X;F(Y;Z))=F(F(X;Y);Z) \tag{5.24}$$

However, certainly from the point of view of applications, one dimensional formal groups are by far the most important, especially one dimensional formal groups over the integers, rings of integers of algebraic number fields, and over polynomial rings over the integers.

The only other that currently seems important is the infinite dimensional formal group $\widehat{W}$ of the Witt vectors which is defined by the same polynomials that define the addition of Witt vectors; see the next subsection 5.2.9.

A standard reference for formal groups is [74].

5.2.8.1 Lazard commutativity theorem

Let A be a ring that has no elements that are simultaneously torsion and nilpotent. Then every one dimensional formal group over A is commutative; i.e. satisfies $F(X,Y)=F(Y,X)$.

5.2.8.2 Universal formal groups

Given a formal group $F(X,Y)$ over A and a morphism of rings $\alpha: A \to B$ one obtains a formal group $\alpha_* F(X,Y)$ over B by applying α to the coefficients of $F(X,Y)$.

A one dimensional commutative formal group $F_L(X,Y)$ over a ring L is called universal[15] if for every one dimensional formal group $F(X,Y)$ over a ring A there is a unique morphism of rings $\alpha^F : L \to A$ such that $\alpha^F_* F_L(X,Y) = F(X,Y)$.

That such a thing exists and is unique is a triviality. What is very remarkable is the theorem of Lazard, [103], that L is the ring of polynomials in an infinity of indeterminates over the integers. The standard proof is awful and highly computational.

[15] This is a rather different ‘universal’ than e.g. in ‘universal enveloping algebra’. The L in these sentences stands for Lazard.

5.2.8.3 Morphisms

A morphism of formal groups from an m-dimensional formal group $F(X;Y)$ to an n-dimensional formal group $G(X;Y)$ is an n-tuple of power series in m indeterminates $\phi(X)$ such that

$$\phi(0)=0,\ G(\phi(X);\phi(Y))=\phi(F(X;Y))$$

If $\phi(X)\equiv X \mod$ (degree 2) the morphism is said to be strict.

5.2.8.4 Logarithms

Let A be a ring of characteristic zero so that the canonical ring morphism $A\to A\otimes_Z Q=A_Q$ is injective; let $F(X,Y)$ be a one dimensional formal group over A. Then over A_Q there exists a power series $f(X)=X+\alpha_2X^2+\cdots$ such that

$$F(X,Y)=f^{-1}(f(X)+f(Y)) \tag{5.25}$$

Here f^{-1} is the compositional inverse of f, i.e. $f^{-1}(f(X))=X$. This f is called the logarithm of F. In the case of the multiplicative formal group, see (5.23), the logarithm is

$$\log(1+X)=X-2^{-1}X^2+3^{-1}X^3-4^{-1}X^4+\cdots$$

Indeed, $\log(1+X+Y+XY)=\log(1+X)+\log(1+Y)$. The terminology derives from this example. The logarithm of a formal group is a strict isomorphism of the formal group to the additive formal group; but over A_Q.

It is at the level of logarithms that the recursive structure of formal groups appears; a recursive structure that was totally unexpected.

There are also logarithms for higher dimensional commutative formal groups.

5.2.8.5 p-typical formal groups

A one dimensional formal group over a characteristic 0 ring is p-typical if its logarithm is of the form

$$f(X)=X+b_1X^p+b_2X^{p^2}+\cdots$$

There is a better definition, see [74], which works always and also in the higher dimensional case. But this one will do for the purposes of the present paper.

Over a $Z_{(p)}$-algebra every formal group is strictly isomorphic to a p-typical one, [30]. If the ring over which the formal group is defined is of characteristic zero the isomorphism is easily described: take the logarithm and change all coefficients of non-p-powers of X to zero.

5.2.8.6 The universal p-typical formal group, [71]

Take a prime number p and consider the following ring with endomorphism

$$\mathbf{Z}[V] = \mathbf{Z}[V_1, V_2, V_3, \cdots],\ \psi(V_n) = V_n^p \tag{5.26}$$

Define

$$\alpha_n(V) = \sum_{i_1+\cdots+i_r=n} p^{-r} V_{i_1} V_{i_2}^{p^{i_1}} V_{i_3}^{p^{i_1+i_2}} \cdots V_{i_r}^{p^{i_1+\cdots+i_{r-1}}} \tag{5.27}$$

Thus the first few of these polynomials are:

$$\begin{aligned}
\alpha_1(V) &= p^{-1}V_1, \\
\alpha_2(V) &= p^{-2}V_1V_1^p + p^{-1}V_2, \\
\alpha_3(V) &= p^{-3}V_1V_1^pV_1^{p^2} + p^{-2}V_1V_2^p + p^{-2}V_2V_1^{p^2} + p^{-1}V_3.
\end{aligned}$$

This sequence of polynomials has both a left and a right recursive structure.

The left recursive structure is

$$\alpha_n(V) = \sum_{i=1}^{n} p^{-1} V_i \psi^i(\alpha_{n-i}(V)) \quad (\text{where } \alpha_0(V) = 1)$$

and the right recursive structure is

$$p\alpha_n(V) = \alpha_{n-1}(V)V_1^{p^{n-1}} + \alpha_{n-2}(V)V_2^{p^{n-1}} + \cdots + \alpha_1(V)V_{n-1}^p + V_n$$

Now consider

$$\begin{aligned}
f_V(X) &= X + \alpha_1(V)X^p + \alpha_2(V)X^{p^2} + \alpha_3(V)X^{p^3} + \cdots \\
F_V(X,Y) &= f_V^{-1}(f_V(X) + f_V(Y))
\end{aligned} \tag{5.28}$$

The left recursive structure is used to prove that $F_V(X,Y)$ is integral, i.e. has its coefficients in $\mathbf{Z}[V]$ and hence is a formal group over $\mathbf{Z}[V]$ and, subsequently, to prove that it is the universal p-typical formal group which means that every p-typical formal group can be obtained from it by a suitable ring morphism from $\mathbf{Z}[V]$.

The right recursive structure then leads to important applications to e.g. complex cobordism theory in algebraic topology and Dirichlet series in number theory.

The important thing here is not that a universal p-typical formal group exists but that it has these very simple and elegant recursive structures.

The universal p-typical formal groups can be simply fitted together to give a construction of the universal formal group.

5.2.8.7 Formal groups from cohomology

Let $h^\star$ be a multiplicative extraordinary cohomology theory with first Chern classes. What all these words really mean is not so important at the present stage. Suffice that many of the better known cohomology theories are like this. The point is that under these circumstances there is a universal formula for the first Chern class of a tensor product of line bundles in terms of the first Chern classes of the factors.

$$c_1(\xi \otimes \eta) = \sum_{i,j} \alpha_{ij} c_1(\xi)^i c_1(\eta)^j$$

defining a formal group over $h^\star(pt)$.

$$F_{h^\star}(X,Y) = \sum \alpha_{ij} X^i Y^j, \quad \alpha_{ij} \in h^\star(pt)$$

Here are some examples.

- $h^\star = H^\star$, ordinary cohomology, $F_H = \widehat{G_\alpha}$, the additive formal group.
- $h^\star = K^\star$, complex K-theory, $F_K(X,Y) = X + Y + uXY$, where u is the Bott periodicity element; a version of the multiplicative formal group.
- $h^\star = MU^\star$, complex cobordism. In this case the formal group has logarithm $f_{MU}(X) = \sum_{n=0}^{\infty} \frac{[CP^n]}{n+1} X^{n+1}$. Here CP^n is n-dimensional complex projective space and $[CP^n]$ is its complex cobordism class in $MU^\star(pt)$. This profound result is due to A S Mishchenko, see appendix 1 of [134].
- $h^\star = BP^\star$, Brown-Peterson cohomology, the 'prime p part' of complex cobordism. Its formal group is the p-typification of the one of complex cobordism, so that its logarithm is $f_{BP}(X) = \sum_{r=0}^{\infty} \frac{[CP^{p^r-1}]}{p^r} X^{p^r}$.

For more details see [74] and the references given there and especially [140].

There is more. The formal group of complex cobordism is the universal one, [138]. This remarkable result is due to Quillen.

The remarkable, elegant and nice aspect here is that in terms of cobordism the universal formal group is so simple and regular.

It follows from the Quillen theorem that $F_{BP}(X,Y)$ with logarithm $f_{BP}(X)$ is the universal p-typical formal group law. But there is also an explicit construction of the universal p-typical formal group law, (5.28). This has all kinds of consequences for complex cobordism and Brown-Peterson cohomology, see [73], [74], [140].

Quillen's theorem also goes a fair way towards establishing that complex cobordism is the most general cohomology theory.

5.2.9 The amazing Witt vectors and their gracious applications

Let **CRing** be the category of unital commutative associative rings. The big Witt vectors constitute a functor W : **CRing** $\to$ **CRing** which has an amazing number of universality properties. For a fair amount of information on this functor see [70] and the references quoted there.

5.2.9.1 Definition of the functor of the big Witt vectors

As a set $W(A) = \Lambda(A)$ is the set of all power series with coefficients in A with constant term 1.

$$W(A) = \Lambda(A) = \{1 + \alpha_1 t + \alpha_2 t^2 + \alpha_3 t^3 + \cdots : \alpha_i \in A\} \tag{5.29}$$

Multiplication of such power series defines an Abelian group structure on $W(A)$ with as neutral element the power series 1. This is the underlying group of the to be defined ring structure on $W(A)$. The multiplication on $W(A)$ is uniquely determined by the requirement that the very special power series $(1-xt)^{-1}$ multiply as

$$(1-xt)^{-1} \star (1-yt)^{-1} = (1-xyt)^{-1} \tag{5.30}$$

and the demands of distributivity (of multiplication over addition) and functoriality. Just how this works out will be indicated immediately below.

The functoriality of $W(-)$ is component-wise, i.e. it is given by

$$W(f)(1 + \alpha_1 t + \alpha_2 t^2 + \alpha_3 t^3 + \cdots) = 1 + f(\alpha_1)t + f(\alpha_2)t^2 + f(\alpha_3)t^3 + \cdots \tag{5.31}$$

The functor W is obviously representable by the ring **Symm** $= Z[h_1, h_2, h_3, \ldots]$ of polynomials in a countable infinity of indeterminates over the integers. The functorial correspondence is:

$$1 + \alpha_1 t + \alpha_2 t^2 + \alpha_3 t^3 + \cdots \leftrightarrow f : Symm \to A, f(h_n) = \alpha_n \tag{5.32}$$

It is convenient to view the h_n as the complete symmetric functions in another countably infinite set of indeterminates $\xi_1, \xi_2, \xi_3, \ldots$ which can be encoded as

$$1 + h_1 t + h_2 t^2 + h_3 t^3 + \cdots = \prod_i \frac{1}{(1-\xi_i t)} \tag{5.33}$$

Now let $h'_1, h'_2, h'_3, \cdots$ be a second set of commuting indeterminates viewed as the complete symmetric functions in $\eta_1, \eta_2, \eta_3, \cdots$ that commute with the ξ. Then distributivity requires that

$$(1 + h_1 t + h_2 t^2 + h_3 t^3 + \cdots) \star (1 + h'_1 t + h'_2 t^2 + h'_3 t^3 + \cdots) = \prod_{i,j} \frac{1}{(1-\xi_i \eta_j t)} \tag{5.34}$$

This makes sense because the right hand side of (5.34) is symmetric in the ξ and in the η and so, by the fundamental symmetric functions theorem there are unique polynomials

$$\Pi_1(h_1; h'_1), \Pi_2(h_1, h_2; h'_1, h'_2), \Pi_3(h_1, h_2, h_3; h'_1, h'_2, h'_3), \cdots \tag{5.35}$$

such that

$$\begin{aligned}&(1+h_1t+h_2t^2+h_3t^3+\cdots)\star(1+h'_1t+h'_2t^2+h'_3t^3+\cdots)=\\&=1+\Pi_1(h_1;h'_1)t+\Pi_2(h_1,h_2;h'_1,h'_2)t^2+\Pi_3(h_1,h_2,h_3;h'_1,h'_2,h'_3)t^3+\cdots\end{aligned}\tag{5.36}$$

(That the multiplication polynomials Π_n depend only on the first n h_i and h'_i is easily seen by degree considerations.)

By functoriality these polynomials determine the multiplication on each $W(A)$ in the sense that for $a(t)=1+a_1t+a_2t^2+a_3t^3+\cdots$ and $b(t)=1+b_1t+b_2t^2+b_3t^3+\cdots$ in $W(A)$ their product is

$$a(t)\star b(t)=1+\Pi_1(a_1;b_1)t+\Pi_2(a_1,a_2;b_1,b_2)t^2+\Pi_3(a_1,a_2,a_3;b_1,b_2,b_3)t^3+\cdots$$

Of course the sum in $W(A)$ is also defined by universal polynomials. These are

$$\Sigma_n(h_1,\cdots,h_n;h'_1,\cdots,h'_n)=\sum_{i+j=n}h_ih'_j \text{ where } h_0=h'_0=1$$

Another way of expressing most of this is to say that

$$\begin{aligned}h_n&\mapsto\Sigma_n(h_1\otimes 1,\cdots,h_n\otimes 1;1\otimes h_1,\cdots,1\otimes h_n)\\h_n&\mapsto\Pi_n(h_1\otimes 1,\cdots,h_n\otimes 1;1\otimes h_1,\cdots,1\otimes h_n)\end{aligned}\tag{5.37}$$

define on **Symm** (most of) the structure of a coring object in the category **CRing**, which hence, via (5.32) defines a functorial ring structure on the $W(A)$.

5.2.9.2 Lambda rings and sigma rings

A pre-sigma-ring (pre-σ-ring) is a unital commutative ring A that comes with extra nonlinear operators that behave (in a very real sense) like symmetric powers. That is, there are operators

$$\sigma_i:A\to A,i=1,2,\cdots;\sigma_1=id\tag{5.38}$$

such that

$$\sigma_n(x+y)=\sigma_n(x)+\sum_{i=1}^{n-1}\sigma_i(x)\sigma_{n-i}(y)+\sigma_n(y)\tag{5.39}$$

It is often useful to have the notation σ_0 for the operator that takes the constant value 1. This notion is equivalent to the better known one of a pre-lambda-ring (pre-λ-ring) but works out just a bit better notationally. The two sets of operations are related by the Wronski-like relations

$$\sum_{i=0}^{n}(-1)^i\sigma_i(x)\lambda_{n-i}(x)=0$$

The lambda operations behave like exterior powers.

Let $\phi:A\to B$ be a morphism in **CRing** and let both A and B carry pre-sigma-ring structures. Then the morphism is said to be a morphism of pre-sigma-rings if it commutes with the sigma operations, i.e. $\phi(\sigma_n^A(x)) = \sigma_n^B(\phi(x))$. A pre-sigma-ring is a sigma ring if the operations satisfy certain universal formulas when iterated and when applied to a product. This is conveniently formulated as follows.

Consider the ring of big Witt vectors $W(A)$ and write an element of it (formally) as

$$\alpha(t) = 1 + \alpha_1 t + \alpha_2 t^2 + \alpha_3 t^3 + \cdots = \prod_i \frac{1}{(1 - \xi_i t)}$$

Then

$$\sigma_n(\alpha(t)) = \prod_{i_1 \leq i_2 \leq \cdots \leq i_n} (1 - \xi_{i_1} \xi_{i_2} \cdots \xi_{i_n} t)^{-1} \tag{5.40}$$

(when written out in terms of the α_i which can be done by the usual symmetric function yoga). This defines a pre-sigma-ring structure on $W(A)$.

A pre-sigma-ring A is a sigma-ring if

$$\sigma_t : A \longrightarrow W(A), x \mapsto 1 + \sigma_1(x)t + \sigma_2(x)t^2 + \sigma_3(x)t^3 + \cdots$$

is a morphism of pre-sigma rings. It is a theorem that $W(A)$ is in fact a sigma-ring. This involves the study of the morphism

$$\sigma^{W(A)} : W(A) \longrightarrow W(W(A)) \tag{5.41}$$

which I like to call the Artin-Hasse exponential.[16]

A ring morphism between sigma-rings is a sigma-ring morphism if it is a morphism of pre-sigma-rings. Let **SigmaRing** be the category of sigma-rings.
Let

$$s_1 : W(A) \longrightarrow A, \alpha(t) \mapsto \alpha_1 \tag{5.42}$$

be the morphism of rings that assigns to a 1-power-series its first coefficient. The Witt vectors now have the following universality property. Let S be a sigma-ring, A a ring and $\phi : S \longrightarrow A$ a morphism of rings, then there is a unique morphism of sigma-rings $\widehat{\phi} : S \longrightarrow W(A)$ such that the following diagram commutes

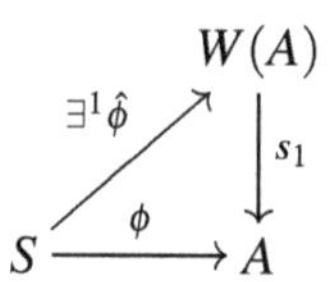

So $W(A) \xrightarrow{s_1} A$ is the cofree sigma-ring over the ring A. Or in other words the functor $W(-)$: **Cring** $\longrightarrow$ **SigmaRing** is right adjoint to the functor the other way that forgets about the sigma structure.

[16] A distant relative of this morphism, viz $W_{p^\infty}(k) \longrightarrow W(W_{p^\infty}(k))$ plays an important role in class field theory. Here k is a finite field and W_{p^∞} is the quotient of the p-adic Witt vectors of the big Witt vectors.

5.2.9.3 The comonad structure on the big Witt vectors

A comonad (also called cotriple) (T,μ,ε) in a category $\mathscr{C}$ is an endo functor T of $\mathscr{C}$ together with a morphism of functors $\mu : T \to TT$ and a morphism of functors $\varepsilon : T \to id$ such that

$$(T\mu)\mu = (\mu T)\mu, \quad (\varepsilon T)\mu = id = (T\varepsilon)\mu \tag{5.43}$$

And a coalgebra for the comonad (T,μ,ε) is an object in the category $\mathscr{C}$ together with a morphism $\sigma : C \to TC$ such that

$$\varepsilon_C\sigma = id, \ (T(\sigma))\sigma = (\mu_{TC})\sigma \tag{5.44}$$

It is now a theorem, [70], that the Artin-Hasse exponential (5.41), which is functorial, together with the functorial morphism (5.42) form a cotriple and that the coalgebras for this cotriple are precisely the sigma-rings.

5.2.9.4 The sigma and lambda ring structures on Symm

Consider

$$\mathbf{Symm} = \mathbf{Z}[h_1,h_2,h_3,\cdots] \subset \mathbf{Z}[\xi_1,\xi_2,\xi_3,\cdots] \tag{5.45}$$

as before. There is a unique sigma-ring structure on $\mathbf{Z}[\xi]$ determined by

$$\sigma_n(\xi_i) = \xi_i^n \tag{5.46}$$

(The corresponding lambda operations are $\lambda_1(\xi_i) = \xi_i, \lambda_n(\xi_i) = 0$ for $n \geq 2$ so that the ξ_i are like line bundles and this is a good way of thinking about them.) The subring **Symm** is stable under these operations and so there is an induced sigma-ring structure on **Symm**.

It is now a theorem that **Symm** with this particular sigma-ring structure is the free sigma-ring on one generator. More precisely:

For every sigma ring S and element $x \in S$ there is a unique morphism of sigma-rings $\mathbf{Symm} \to S$ that takes h_1 into x.

The universality properties described in subsections 5.2.9.2, 5.2.9.3, 5.2.9.4 are far from unrelated; see section 5.3.3 below. A totally different universality property of the Witt vectors is the following one.

5.2.9.5 Cartier's first theorem

The (infinite dimensional) formal group of the Witt vectors 'is' the sequence of addition polynomials $\Sigma_1,\Sigma_2,\ldots$ in $X_1,X_2,\cdots;Y_1,Y_2,\cdots$. This formal group is denoted $\widehat{W}$. A fourth universality property of the Witt vectors holds in this setting.

Given two formal groups F and G of dimensions m and n respectively a morphism of formal groups $\alpha : F \to G$ is an n-tuple of power series with zero constant terms $\alpha_1, \cdots, \alpha_n$ in m variables such that

$$G(\alpha_1(X), \cdots, \alpha_n(X); \alpha_1(Y), \cdots, \alpha_n(Y)) = (\alpha_1(F(X,Y)), \cdots, \alpha_n(F(X,Y))) \tag{5.47}$$

A curve in an n-dimensional formal group F is simply an n-tuple of power series in one variable, say, t. In $\widehat{W}$ consider the particular curve $\gamma_0(t) = (t, 0, 0, \cdots)$. Then Cartier's first theorem says that for every formal group F and curve $\gamma(t)$ in it, there is a unique morphism of formal groups $\widehat{W} \to F$ that takes $\gamma_0(t)$ into $\gamma(t)$.

5.2.10 *The star example:* Symm

Here is a list of most of the objects with which this subsection will be concerned. Those which have not already been defined above will be described in section 5.3.3 below.

- **Symm** $= \mathbf{Z}[h_1, h_2, \cdots] = \mathbf{Z}[c_1, c_2, \cdots] \subset \mathbf{Z}[\xi_1, \xi_2, \cdots]$, the ring of symmetric functions in an infinity of indeterminates. Here h_n is the n-th complete symmetric function in the ξ's and the c_n stand for the elementary symmetric functions. I am writing c_n rather than e_n because in the present context the c_n will correspond to Chern classes
- $U(\Lambda)$, the universal lambda ring on one generator
- $R(W)$, the representing ring of the functor of the big Witt vectors; see subsection 5.2.9 above
- $R(S) = \bigoplus_{n=0}^{\infty} R(S_n)$, the direct sum of the rings of (the Grothendieck groups of) complex representations of the symmetric groups with the so-called exterior product; if ρ is a representation of S_r and σ is a representation of the symmetric group on s letters S_s then $\rho\sigma = Ind_{S_r \times S_s}^{S_{r+s}}(\rho \otimes \sigma)$. By decree $R(S_0)$ is equal to $\mathbf{Z}$. There is also a comultiplication: if σ is a representation of S_n $\mu(\sigma) = \sum_{r+s=n} Res_{S_r \times S_s}^{S_n}(\sigma)$. Together with obvious unit and counit morphisms this defines a Hopf algebra. (The antipode comes for free because of the graded connected situation.)
- $R_{rat}(GL_\infty)$, the (Grothendieck) ring of rational representations of the infinite linear group
- $E(\mathbf{Z})$, the value of the exponential functor from [81] on the ring of integers
- $U(\widehat{W})$, the covariant bialgebra of the formal group of the Witt vectors
- $H^\star(\mathbf{BU}; \mathbf{Z})$, the cohomology of the classifying space of complex vector bundles, **BU**
- $H_\star(\mathbf{BU}; \mathbf{Z})$, the homology of the classifying space **BU**

These are all isomorphic and that implies that **Symm** is very rich in structure indeed. Nor is that all. For instance each of the components $R(S_n)$ of $R(S)$ is a lambda ring in its own right (inner plethysm). Further the functor of the big Witt vectors is lambda ring valued. However, this paper is not about **Symm** and its extraordinarily rich structure,[17] but about niceness results. That includes 'nice proofs'. That is proofs of isomorphism between all these objects that derive from their universality, (co)freeness, ... properties and rely minimally on calculations. To what extent there are currently such proofs will be discussed below in subsection 5.3.3.

Two more objects that fit in this picture are the rational Witt vector functor in its role in the K-theory of endomorphisms, [4], and the Grothendieck group $K(\mathbf{P}_A)$ of polynomial functors $\mathbf{Mod}_A \to \mathbf{Mod}_k$, where A is an algebra over a field k and $\mathbf{Mod}_A$ is the category of right A-modules, [117]. If $A = k$ this object is again isomorphic to the nine objects listed above.

The various isomorphisms and relations concerning which I think I have something to say are depicted in the diagram below.

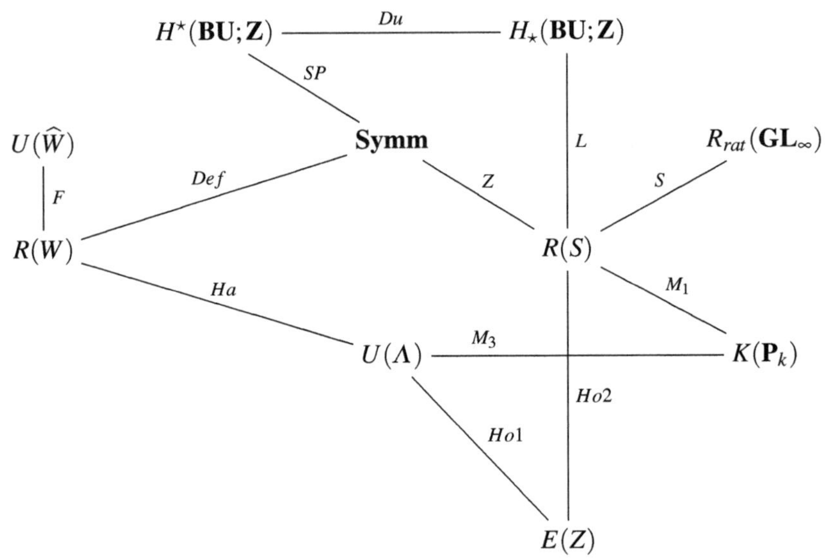

The bottom object here, viz $E(\mathbf{Z})$, has not yet been described in any way. It is again defined by an adjoint functor situation and, again, it is one which picks up extra structure. It will be described and discussed briefly in section 5.3.3 below.

Also it seems from the diagram that the Hopf algebra $R(S) = \bigoplus_{n=0}^{\infty} R(S_n)$ is the central object rather than **Symm**.

[17] I plan a future paper on that; meanwhile see [70].

5.2.11 Product formulas

The simplest (arithmetic) product formula concerns the real and p-adic absolute values of a rational number

$$|\alpha|_\infty = \prod_p |\alpha|_p^{-1}$$

where the product on the right is over all prime numbers p. There are more formulas of this type. This leads to a view of things that is expressed as follows by Yuri Manin in [119], Reflections on arithmetical physics, pp 149ff.

"Now we can see the following pattern

- (at least some) essential notions of real and complex calculus and geometry have their adèlic counterparts;
- adèlic objects have a strong tendency to be simpler than their Archimedean components, e.g. the adèlic fundamental domains of arithmetical discrete subgroups of semisimple groups usually have volume 1 (the Siegel-Tamagawa-Weil philosophy ...);
- due to this fact and to product formulas like (2) or (3) embodying the idea of democracy for all topologies, information on the real component of an adèlic object can be read off either from the real component or the product of the p-adic components for all p's.

With some strain one can generalize and state the following principle which is the main conjecture of this talk.

On the fundamental level our world is neither real, nor p-adic; it is adèlic. For some reasons reflecting the physical nature of our kind of living matter (e.g., the fact that we are built of massive particles), we tend to project the adèlic picture onto its real side. We can equally well spiritually project it upon its non-Archimedean side and calculate most important things arithmetically."

There are applications of this idea to the Polyakov measure (Polyakov partition function), loc. cit., string theory, [58], Yang-Mills theory, [6], and much more, see, for a start, (the bibliography of) [95]. Add to this that the p-adic versions are often easier to handle and one finds some good justification for the discipline of p-adic physics.

5.3 Some first results and theorems

5.3.1 Freeness theorems

The only general freeness theorem that I know about is the one from [57]. This one says that cogroups (cogroup objects) in the category of algebras over an operad are free. This covers for instance one of the Kan results, the Leray theorem, the Milnor–

Moore theorem and probably several more. At this stage it is unclear how far it goes.

I don't think it can be made to take care of the subobject freeness theorems; but there probably is a general theorem, yet to be formulated and proved, that can take care of those.

5.3.2 On the Lazard universal formal group theorem

The Lazard universal formal group theorem says that there exists a universal (one dimensional) commutative formal group (trivial) and that the underlying ring is free commutative polynomial in an infinity of indeterminates (surprising and far from trivial). The standard proof is long, laborious, and computational, even when simplified and streamlined as in [60], see also [140].

Having a candidate universal formal group available, as in [72], [74] helps a great deal, see [74], pp 27–30. But the proof is still mainly computational; also the construction of the candidate universal formal group involves choices of coefficients, which mars things. One dreams of a proof which mainly relies on universality properties.

In this connection there is a rather different proof due to Cristian Lenart, [108], which seems to have promising aspects. One ingredient, which I consider promising, is the following. Consider the power series

$$f_b(X) = X + b_2X^2 + b_3X^3 + \cdots$$

over $\mathbf{Z}[b]$. Here the b's are indeterminates. Now form

$$F_b(X,Y) = f_b(f_b^{-1}(X) + f_b^{-1}(Y))$$

This is of course a formal group over $\mathbf{Z}[b]$. It is proved[18] in loc. cit. that the coefficients of $F_b(X,Y)$ generate a free polynomial subring, L, of $\mathbf{Z}[b]$ and that regarded as a formal group over the subring L $F_b(X,Y)$ is universal. Of course L is truly smaller than $\mathbf{Z}[b]$. To start with $2b_2 \in L$, but $b_2 \notin L$.

This next bit is pure speculation. The first Cartier theorem on formal groups says that the formal group of Witt vectors, $\widehat{W}$, represents the functor 'curves'. This is a rather different universality property for formal groups. The covariant bialgebra of $\widehat{W}$ is **Symm**. One wonders whether this can be used to prove the Lazard theorem.

[18] The result is nice; I consider the proof highly unsatisfactory.

5.3.3 *Objects and isomorphisms in connection with* Symm

This whole subsection is concerned with the objects and isomorphisms in the diagram at the end of section 5.2.10.

5.3.3.1 The isomorphism 'Ha' between $R(W)$, the representing ring of the functor of the big Witt vectors and $U(\Lambda)$, the free lambda ring on one generator

Here is a synopsis of the relevant bits of structure. The ring $R(W)$ represents a (covariant) functor that carries a comonad structure, and the coalgebras for this comonad are precisely the lambda rings. That is all that is needed.

Let $\mathscr{C}$ be a category and let (T, μ, ε) be a comonad in $\mathscr{C}$. Now let $(Z, z \in T(Z))$ represent the functor T. That is, there is a functorial bijection $\mathscr{C}(Z,A) \to T(A), f \mapsto T(f)(z)$. The comonad structure gives in particular a morphism $\sigma : T \to TZ$, viz the image of id_Z under $\mu_Z : T(Z) = \mathscr{C}(Z,Z) \to T(T(Z)) = \mathscr{C}(Z,T(Z))$. This defines a 'coalgebra for T' structure on Z. Now let (A,σ) be a coalgebra for the comonad T and let α be an element of A. Consider the element $\sigma(\alpha) \in T(A) = \mathscr{C}(Z,A)$. This gives a unique morphism of T-coalgebras that takes z into α. There are of course a number of things to verify both at this categorical level and to check that these categorical considerations fit with the explicit constructions carried out in the previous subsections. This is straightforward.

Thus the isomorphism 'Ha' is a special case of a quite general theorem and the proof uses no special properties but only universal and other categorical notions. This is the kind of proof I would like to have for all the isomorphisms in the diagram.

5.3.3.2 The isomorphism 'Z' between $R(S)$ and Symm

This is handled by the Zelevinsky theorem, [164] and [69], chapter 3. The Zelevinsky theorem deals with PSH algebras (over the integers). The acronym 'PSH' stands for 'Positive-Selfadjoint-Hopf'. Actually it is about (nontrivial) graded connected positive self-adjoint Hopf algebras with a distinguished (preferred) homogenous basis. The Hopf algebra is also supposed to be of finite type so that each homogenous component is a free Abelian group of finite rank.

An inner product is defined by declaring this basis to be orthonormal. The positive elements of the Hopf algebra are the nonnegative (integer coefficient) linear combinations of the distinguished basis elements. Let m and μ denote the multiplication and comultiplication respectively.

Selfadjoint (selfdual) now means

$$\langle m(x \otimes y), z \rangle = \langle x \otimes y, \mu(z) \rangle$$

and positivity means that if the elements of the distinguished basis are denoted by ω_i etc., and

$$m(\omega_i \otimes \omega_j) = \sum_r \alpha_{i,j}^r \omega_r, \quad \mu(\omega_i) = \sum_{r,s} b_i^{r,s} \omega_r \otimes \omega_s$$

then $\alpha_{i,j}^r \geq 0$ and $b_i^{r,s} \geq 0$. Suppose now that there is precisely one among the distinguished basis elements that is primitive,[19] then (the main part of) the Zelevinsky theorem says that the Hopf algebra in question is isomorphic (as a Hopf algebra) to **Symm**, possibly degree shifted.

An example of a PSH algebra is $R(S)$:

- The distinguished basis is formed by the irreducible representations of the various S_n
- The positive elements are the real (as opposed to virtual) representations, and so multiplication and comultiplication are positive.
- The selfadjointness comes from Frobenius reciprocity
- The Hopf property is handled by (a consequence of) the Mackey double coset theorem.

Using the isomorphism all structure can be transferred making **Symm** also a PSH algebra. An odd thing is that this is not proved directly. The distinguished basis turns out to be formed by the Schur functions. The problem is positivity. There seems to be no direct proof in the literature that the product of two Schur functions is a nonnegative linear combination of Schur functions.

I used to think that this theorem did not count in the context of the diagram because it uses such seemingly non-algebraic things as positivity and distinguished basis. However in the setting of $R(S)$ these are, see above, entirely natural.

There is one more thing I would like to say in this context. The fourth and final step of the proof of the Zelevinsky theorem (in the presentation of [69]) essential use is made of something called the Bernstein morphism. This is a morphism

$$H \to H \otimes \mathbf{Symm}$$

defined for any commutative associative graded connected Hopf algebra H. If one takes $H = \mathbf{Symm}$ this is precisely the morphism that defines the multiplication on the big Witt vectors. This is a "coincidence" that cries out for further investigation. For a completely different way of establishing that **Symm** and $R(S)$ are isomorphic see [7]. For still another and very elegant proof of this result see [111], [112]. It seems that the theorem actually goes back to Frobenius, [59].

5.3.3.3 The isomorphism 'S' from $R(S)$ to $R_{rat}(\mathbf{GL}_\infty)$

This is Schur-Weyl duality which has its origins in Schur's thesis of 1901, [143]. The subject of Schur-Weyl duality has by now evolved into what is practically a

[19] There is always at least one because of graded connectedness (and nontriviality).

small specialism of its own. A search in the MathSciNet and ZMATH databases gives numerous results. There are quantum and super versions and there are interrelations with such diverse fields as quantum and statistical mechanics, tilting theory, combinatorics, random walks on unitary groups, ... A selection of references is [16], [43], [45], [61], [63], [64], [65], [83], [96], [110], [143], [144], [161], [163], [7], [117], [49].

Here is what is probably the simplest incarnation of Schur-Weyl duality. Let V be a finite dimensional vector space over a field of characteristic 0. Form the n-th tensor product

$$T^n(V) = V \otimes \cdots \otimes V$$

The symmetric group S_n acts on this by permuting the factors, which gives a finite dimensional representation of S_n that can be decomposed into its isotypic components

$$T^n(V) = \bigoplus_{\pi} \mathrm{Hom}_{kS_n}(E_\pi, T^n(V) \otimes E_\pi) = \bigoplus_{\pi} F_\pi(V) \otimes E_\pi \tag{5.48}$$

functorially in V. Here the E_π are the distinct irreducible kS_n modules. If now $A : V \to V$ is a linear transformation $F_\pi(A) : F_\pi(V) \to F_\pi(V)$ is an 'invariant matrix' in the sense of Schur, [143]. This is taken from [117].

Taking invertible A one obtains a representation $F_\pi(V)$ of $\mathbf{GL}(V)$. This can also be seen as coming from the action of $\mathbf{GL}(V)$ on $T^n(V)$ defined by $g(v_1 \otimes \cdots \otimes v_n) = gv_1 \otimes \cdots \otimes gv_n$, noting that this action commutes with the S_n action on $T^n(V)$ and using the double commutant theorem.

The middle term in (5.48) makes it clear that this is some kind of duality. What I would really like is to have is a pairing $R(S) \times R_{rat}(\mathbf{GL}) \to \mathbf{Z}$ defined directly, which then gives this duality. At the 'finite level' described above this can probably be done by looking at the trace form $\langle X, Y \rangle = \mathrm{Trace}(XY)$ on $\mathrm{End}(T^n(V))$, [64], section 9.1; [163], page 19. But not, it seems, without bringing in a lot of representation theory.

5.3.3.4 On a possible isomorphism 'L' between $R(S)$ and $H_\star(\mathbf{BU};\mathbf{Z})$

This is mostly speculative. First both rings (as Abelian groups) have a natural basis indexed by partitions. Second there is a bit of positive evidence in [79], where in section 11 a (nontrivial, i.e. with jumps) family of representations is constructed of S_{n+m} that is parametrized by the Grassmann manifold $\mathbf{Gr}_n(\mathbf{C}^{n+m})$.

5.3.3.5 On the isomorphism 'Du' between $H^\star(\mathbf{BU};\mathbf{Z})$ and $H_\star(\mathbf{BU};\mathbf{Z})$

This is a matter of homology–cohomology duality for oriented manifolds. Plus autoduality of the Hopf algebras involved. (Both carry natural Hopf algebra structures.)

5.3.3.6 On the isomorphism 'SP' between $H^\star(\mathbf{BU};\mathbf{Z})$ and Symm

First one defines Chern classes, for instance as in [128], chapter 14; the definition of the first Chern class that is in [94] is one I particularly like.

The i-th Chern class associates to a complex vector bundle V over a suitable space X an element $c_i(V)$ of the cohomology group $H^{2i}(X;\mathbf{Z})$. One of the more important properties of the Chern classes is 'functoriality'. Let $f: Y \to X$ be continuous and let $f^\star V$ be the vector bundle pullback of V. Then

$$c_i(f^\star V) = f^\star(c_i(V))$$

(The notation is a bit unfortunate in that there are two different $f^\star$ in the formula; but is traditional). A second important property is the 'Whitney sum formula'. Let

$$c(V) = 1 + c_1(V) + c_2(V) + \cdots$$

be the so-called total Chern class (also sometimes called complete Chern class). Let W be a second complex vector bundle over X. Then

$$c(V \oplus W) = c(V)c(W)$$

where on the right hand side the cohomology cup product is used. And in fact together with $c_0(V) = 1$ and a normalization condition that specifies the total Chern class of the canonical (tautological) line bundle over the complex projective spaces $\mathbf{Gr}_1(\mathbf{C}^n)$ these two properties completely determine the Chern classes. See also [68], theorem 3.2 on page 78.

Next one calculates the cohomology of the classifying spaces $\mathbf{BU}_n$ to be

$$H^\star(\mathbf{BU}_n;\mathbf{Z}) = \mathbf{Z}[c_1, c_2, \cdots, c_n]$$

where the c_i are the Chern classes of the canonical vector bundle γ_n over $\mathbf{BU}_n$. For instance with induction starting with the very simple case $\mathbf{BU}_1 = \mathbf{CP}^\infty$ which has a CW complex cell decomposition with precisely one cell in every even dimension. This is the way it is done in [128]. One can also use special sequences. It follows that

$$H^\star(\mathbf{BU};\mathbf{Z}) = \mathbf{Z}[c_1, c_2, \cdots, c_n, \cdots]$$

which is isomorphic, at least as rings, to **Symm**. This is precisely the kind of calculatory proof that I do not like.

However, there is the following aspect. It is often a good idea to view **Symm** as the symmetric functions in an infinity of indeterminates

$$\mathbf{Symm} \subset \mathbf{Z}[\xi_1, \xi_2, \xi_3, \cdots]$$

Now on the topological side consider the canonical line bundle $\gamma_1 \to \mathbf{BU}_1$ and take the n-fold product $\gamma_1 \times \cdots \times \gamma_1$. This an n-dimensional bundle over the n-fold product $\mathbf{BU}_1 \times \cdots \times \mathbf{BU}_1$. The cohomology of this space is

$$\mathbf{Z}[\eta_1, \cdots, \eta_n]$$

where η_i is the first Chern class of the i-th γ_i. Also by the Whitney sum formula

$$c(\gamma_1 \times \cdots \times \gamma_1) = (1+\eta_1)\cdots(1+\eta_n)$$

Now by the classifying space property of $\mathbf{BU}_n$ there is a homotopy class of maps $f : \mathbf{BU}_1 \times \cdots \times \mathbf{BU}_1 \to \mathbf{BU}_n$ such that the pullback of γ_n by f is $\gamma_1 \times \cdots \times \gamma_1$. Using functoriality it follows that $f^\star$ takes $c_i \in H^\star(\mathbf{BU}_n; \mathbf{Z})$ to the i-th elementary symmetric function in the η's and that $H^\star(\mathbf{BU}_n; \mathbf{Z})$ manifests itself as the ring of symmetric functions in $Z[\eta_1, \cdots, \eta_n]$. This is taken from page 189 of [128];

Add to this that the Chern classes of the γ_n (the universal Chern classes) can be described explicitly in terms of Schubert cycles, and, possibly, this can be worked up to a much less calculatory proof of the isomorphism 'SP'.

5.3.3.7 The isomorphism 'F' between $R(W)$ and $U(\widehat{W})$

Consider an n-dimensional formal group F over a (unital commutative associative) ring A. Here n can be infinity. It is given by n power series in $2n$ indeterminates grouped in two groups of n indeterminates with coefficients in A. Let $R(F)$ be the ring of power series over A in n indeterminates. Then the n power series of the formal group F define a bialgebra like structure

$$R(F) \to R(F)\widehat{\otimes}R(F)$$

This object is called the contravariant bialgebra of the formal group. (It is really needed (in general) to take the completed tensor product; even for $n = 1$ one has $A[[X]] \otimes A[[Y]] \subsetneq A[[X,Y]]$.)

$R(F)$ is given the usual power series topology. Now form

$$U(F) = \mathbf{Mod}_{A,cont}(R(F), A) \tag{5.49}$$

This is the covariant bialgebra (in fact Hopf algebra) of the formal group F. Inversely one can obtain $R(F)$ from $U(F)$; just how will not be needed here.

In the case of the formal group $\widehat{W}$ of the Witt vectors (over the integers) the power series defining it are in fact polynomials. And thus the restriction to

$$R(W) = \mathbf{Z}[X_1, X_2, \cdots] \subset \mathbf{Z}[[X_1, X_2, \cdots]] = R(\widehat{W})$$

of $R(\widehat{W}) \to R(\widehat{W})\widehat{\otimes}R(\widehat{W})$ lands in $R(W) \otimes R(W)$. As the polynomials are dense in the power series, in this polynomial case, formula (5.49) is equivalent to

$$U(\widehat{W}) = \mathbf{Mod}_{\mathbf{Z}}(R(W), \mathbf{Z})$$

and thus isomorphism 'F' is a consequence of the autoduality of $R(W) = \mathbf{Symm}$.

5.3.3.8 On the isomorphisms 'M1' and 'M3' between $R(S)$, $K(\mathbf{P}_k)$ and $U(\Lambda)$

One sees from formula (5.48) in subsection 5.3.3.3 that each irreducible representation of S_n defines a functor of $\mathbf{V}_k$ to itself that is polynomial. Here $\mathbf{V}_k$ is the category of finite dimensional vector spaces over the field k, and polynomial means that for each pair of vector spaces U,V the mapping $F : Hom(U,V) \to Hom(F(U),F(V))$ is polynomial. Let now $\mathbf{P}_k$ be the category of polynomial functors $\mathbf{V}_k \to \mathbf{V}_k$ of bounded degree and $K(\mathbf{P}_k)$ its Grothendieck group. Then the remarks just made practically establish the isomorphism 'M1'.

Next, $K(\mathbf{P}_k)$ carries a λ-ring structure induced by composition with the exterior powers $\Lambda^i : \mathbf{V}_k \to \mathbf{V}_k$. It turns out that it thus becomes the free λ-ring on one generator, [7], [118]. This is 'M3'.

It needs to be sorted out whether the composition of 'M1' and 'M3' equals the composition of 'Z' and 'Ha'.

The main aim of [118] is to generalize this in various ways. Let A be a k algebra, $\mathbf{V}_A$ the category of finitely generated projective left A modules, $\mathbf{P}_A$ the category of polynomial functors $\mathbf{V}_A \to \mathbf{V}_k$ of bounded degree and $K(\mathbf{P}_A)$ its Grothendieck group. Then $K(\mathbf{P}_k)$ is the free λ-ring generated by the classes of the functors $P \mapsto E \otimes_A P$ where E runs through a complete set of non-isomorphic finite dimensional simple right A-modules.

When applied to the group ring of a finite group there is also the result that $\bigoplus_{n\geq 0} R(G \sim S_n)$ is the free λ-ring on the irreducible representation of G. (Here $G \sim S_n$ is the wreath product of G and S_n.) Thus 'M1' and 'M3' are just the simplest cases of much more general results, which makes them nicer in my view.

5.3.3.9 On the object $E(\mathbf{Z})$ and the isomorphisms 'Ho1' and 'Ho2'

Peter Hoffman noted that there is a nice functor E, denoted 'exp' in [81] that makes some of what went before more elegant.

Let **Ab** be the category of Abelian groups and **GrRing** that of (unital ungraded–commutative) graded rings. An object R of **GrRing** is a direct sum of Abelian groups R_i together with multiplications $R_i \otimes R_j \to R_{i+j}$ making $\oplus_i R_i$ a unital commutative ring. As in the case of the big Witt vectors one considers the "1-units". To be precise consider the functor

$$\hat{\ } : \mathbf{GrRing} \to \mathbf{Ab} \text{ defined by } \widehat{R} = 1 + \prod_{i=1}^{\infty} R_i \tag{5.50}$$

where the Abelian group structure is given by multiplication.

Note that the functor of the big Witt vectors is given by $S \mapsto S[[t]] \mapsto S[[t]]\hat{\ }$. What this means is completely unexplored.

The functor (5.50) has a left adjoint $\mathbf{Ab} \to \mathbf{GrRing}$, here denoted E, so that there is the functorial equality

$$\mathbf{GrRing}(E(A),R) = \mathbf{Ab}(A,\widehat{R})$$

As a left adjoint $E(A)$ should be thought of as some kind of free object and, as is so often the case with functors that are part of an adjunction it picks up all kinds of extra structure. In this case it is first of all a Hopf algebra (as happened with the universal enveloping algebra). This comes from the observation that

$$E(A\oplus B) = E(A)\otimes E(B)^{20}$$

$E(A)$ carries a natural λ-ring structure. (Though I find the construction very difficult and, frankly, definitely on the ugly side.)

However, it is worth exploring further as it goes through the notion of what the author calls an ω-ring, a notion equivalent to that of a λ-ring but whose axioms only involve linear maps. This gives one a shot at solving a rather vexing matter. **Symm** is a λ-ring; it is also selfdual. So, morally speaking, there should be something like a 'dual λ-ring structure' on it.

Returning to the paper [81], the main theorems appear to be

$$\bigoplus_{n\geq 0} R(G\sim S_n) \cong E(R(G))$$

$$E(A) \text{ is the free } \lambda\text{-ring generated by } A$$

which are very nice results showing that the functor E merits further attention.

5.3.3.10 The K-theory of endomorphisms

Let A be a unital commutative ring. Consider the category $\text{End}(A)$ of pairs (P,f) where P is a finitely generated projective A-module and f an endomorphism of P. A morphism $\phi : (P,f) \to (Q,g)$ in $\text{End}(A)$ is a morphism ϕ of A-modules that commutes with the given endomorphisms, i.e. $g\phi = \phi f$. There is an obvious notion of exact sequence in $\text{End}(A)$ and so one can form the Grothendieck group and ring,[21] $K(\text{End}(A))$, the study of which was initiated by Gert Almkvist, [3], [4].

Given $(P,f)\in \text{End}(A)$ let Q be a finitely generated module such that $P\oplus Q$ is free and consider the endomorphism $f\oplus 0$ of this module and its characteristic polynomial $\det(1+t(f\oplus 0))$. This is a polynomial in t that does not depend on Q. This induces a homomorphism $K(\text{End}(A)) \to W(A)$, where $W(-)$ is the functor of the big Witt vectors, that is obviously zero on $K(A)$. (The projective modules over A are imbedded in $\text{End}(A)$ as pairs $(A,0)$). Thus there results a morphism (of rings in fact)

$$c : K(\text{End}(A))/K(A) = W_0(A) \longrightarrow W(A)$$

functorial in A. Almkvist now proves:

[20] This formula also illustrates that 'exponential' or 'exp' is a most apt appellation.

[21] The multiplication is induced by the tensor product.

The morphism c is injective for all A and the image of c (for a given A) consists of all power series $1+\alpha_1 t+\alpha_2 t^2+\cdots$ that can be written in the form

$$1+\alpha_1 t+\alpha_2 t^2+\cdots = \frac{1+b_1 t+b_2 t^2+\cdots+b_r t^r}{1+d_1 t+d_2 t^2+\cdots+d_n t^n} \text{ with } b_i, d_j \in A$$

For obvious reasons I call these rational Witt vectors. A first question is now whether this functor $W_0(-)$ is representable. It is, [75]. This requires some preparation. Consider the ring

$$\mathbf{Z}[X] = \mathbf{Z}[X_1, X_2, X_3, \cdots]$$

of polynomials in a countable infinity of commuting indeterminates. Form the Hankel matrix

$$\begin{pmatrix} 1 & X_1 & X_2 & X_3 & \cdots \\ X_1 & X_2 & X_3 & X_4 & \cdots \\ X_2 & X_3 & X_4 & X_5 & \cdots \\ \vdots & \vdots & \vdots & \vdots & \ddots \end{pmatrix}$$

Now let J_n be the ideal in $\mathbf{Z}[X]$ generated by all $(n+1)\times(n+1)$ minors of this Hankel matrix. These ideals define a topology on $\mathbf{Z}[X]$ which for the present purposes I will call the J-topology. The representability result is now as follows.

For each rational Witt vector $\alpha(t) = 1+\alpha_1 t+\alpha_2 t^2+\cdots \in W_0(A)$ let $\phi_{\alpha(t)}$: $\mathbf{Z}[X] \to A$ be the ring morphism defined by $X_i \mapsto \alpha_i$. Then $\alpha(t) \mapsto \phi_{\alpha(t)}$ is a functorial and injective morphism from $W_0(A)$ to ring morphisms $\mathbf{Z}[X] \to A$ that are continuous with respect to the J-topology on $\mathbf{Z}[X]$ and the discrete topology on A. If A is Fatou, so in particular if A is integral and Noetherian, the correspondence is bijective.

Here Fatou is a technical condition that is of no particular importance for this paper. Suffice it to say that a Noetherian integral domain is Fatou. Incidentally the quotient rings $\mathbf{Z}[X]/J_n$ are integral domains, but they are not Noetherian and not Fatou.

For a host of other results, including a determination of the operations in the K-theory of endomorphisms, see [3], [4], [75].

5.3.3.11 Leftovers

- **Symm** is an object with an enormous amount of compatible structure: Hopf algebra, inner product, selfdual (as a Hopf algebra), PSH, coring object in the category of rings, ring object in the category of corings (up to a little bit of unit trouble), Frobenius and Verschiebung endomorphisms, free algebra on the cofree coalgebra over Z (and the dual of this: cofree coalgebra over the free algebra on one element), several levels of lambda ring structure, ...
  The question arises which ones of these have natural interpretations in the other nine incarnations occurring in the diagram (and whether the isomorphisms indicated are the right ones for preserving these structures).

- **Symm** represents the functor of the big Witt vectors $W(A) = \{1 + \alpha_1 t + \alpha_2 t^2 + \cdots : \alpha_i \in A\}$. Now $\mathbf{Hopf}(\mathbf{Symm}, \mathbf{Symm}) = W(\mathbf{Z})$, [112]. This comes about because on the one hand **Symm** is the free algebra on the cofree coalgebra over **Z**, and on the other the cofree coalgebra over the free algebra over **Z**.
  This is a curiosity that certainly merits some thought and one wonders whether something similar occurs elsewhere.

The list of references below contains more items than are actually referred to in the text above. The others are included because I know or suspect that there are more niceness results in them.

References

1. J. W. Alexander. An Example of a Simply Connected Surface Bounding a Region which is not Simply Connected. *Proc. Nat. Acad. Sci.*, 10:8–10, 1924.
2. C. Allday and V. Puppe. *Cohomological methods in transformation groups*, volume 32 of *Cambridge Studies in Advanced Mathematics*. Cambridge University Press, Cambridge, 1993.
3. Gert Almkvist. The Grothendieck ring of the category of endomorphisms. *J. Algebra*, 28:375–388, 1974.
4. Gert Almkvist. K-theory of endomorphisms. *J. Algebra*, 55(2):308–340, 1978.
5. David J. Anick Hopf algebras up to homotopy. *J. Amer. Math. Soc.*, 2(3):417–453, 1989.
6. Aravind Asok, Brent Doran, and Frances Kirwan. Yang-Mills theory and Tamagawa numbers: the fascination of unexpected links in mathematics. *Bull. Lond. Math. Soc.*, 40(4):533–567, 2008.
7. M. F. Atiyah. Power operations in K-theory. *Quart. J. Math. Oxford Ser. (2)*, 17:165–193, 1966.
8. M. F. Atiyah and D. O. Tall. Group representations, λ-rings and the J-homomorphism. *Topology*, 8:253–297, 1969.
9. P. C. Baayen. *Universal morphisms*, volume 9 of *Mathematical Centre Tracts*. Mathematisch Centrum, Amsterdam, 1964.
10. Stefan Banach and Alfred Tarski. Sur la décomposition des ensembles de points en parties respectivement congruentes. *Fundamenta Mathematicae*, 6:244–277, 1924. Available from the Polish Virtual Library of Science http://matwbn.icm.edu.pl/.
11. Michael F. Barnsley, Robert L. Devaney, Benoit B. Mandelbrot, Heinz-Otto Peitgen, Dietmar Saupe, and Richard F. Voss. *The science of fractal images*. Springer-Verlag, New York, 1988. With contributions by Yuval Fisher and Michael McGuire.
12. Hyman Bass. Big projective modules are free. *Illinois J. Math.*, 7:24–31, 1963.
13. Hyman Bass. Projective modules over free groups are free. *J. Algebra*, 1:367–373, 1964.
14. Hans Joachim Baues. *Algebraic homotopy*, volume 15 of *Cambridge Studies in Advanced Mathematics*. Cambridge University Press, Cambridge, 1989.
15. Alan F. Beardon. *Iteration of rational functions*, volume 132 of *Graduate Texts in Mathematics*. Springer-Verlag, New York, 1991. Complex analytic dynamical systems.
16. Georgia Benkart and Sarah Witherspoon. Representations of two-parameter quantum groups and Schur-Weyl duality. In *Hopf algebras*, volume 237 of *Lecture Notes in Pure and Appl. Math.*, pages 65–92. Dekker, New York, 2004.
17. George M. Bergman. Centralizers in free associative algebras. *Trans. Amer. Math. Soc.*, 137:327–344, 1969.
18. Israel Berstein. On co-groups in the category of graded algebras. *Trans. Amer. Math. Soc.*, 115:257–269, 1965.

19. Czesław Bessaga. Every infinite-dimensional Hilbert space is diffeomorphic with its unit sphere. *Bull. Acad. Polon. Sci. Sér. Sci. Math. Astronom. Phys.*, 14:27–31, 1966.
20. Czesław Bessaga and Aleksander Pełczyński. *Selected topics in infinite-dimensional topology*. PWN—Polish Scientific Publishers, Warsaw, 1975. Mathematical Monographs, Vol. 58.
21. Francis Borceux. *Handbook of categorical algebra. 1. Basic category theory*, volume 50 of *Encyclopedia of Mathematics and its Applications*. Cambridge University Press, Cambridge, 1994. Chapter 3: Adjoint functors.
22. Armand Borel. Sur la cohomologie des espaces fibrés principaux et des espaces homogènes de groupes de Lie compacts. *Ann. of Math. (2)*, 57:115–207, 1953.
23. Armand Borel. Topology of Lie groups and characteristic classes. *Bull. Amer. Math. Soc.*, 61:397–432, 1955.
24. Raoul Bott and Hans Samelson. On the Pontryagin product in spaces of paths. *Comment. Math. Helv.*, 27:320–337 (1954), 1953.
25. R. M. Bryant and Marianne Johnson. Lie powers and Witt vectors. *J. Algebraic Combin.*, 28(1):169–187, 2008.
26. Victor Buchstaber and Andrey Lazarev. Dieudonné modules and p-divisible groups associated with Morava K-theory of Eilenberg-Mac Lane spaces. *Algebr. Geom. Topol.*, 7:529–564, 2007.
27. Daniel Bump. *Lie groups*, volume 225 of *Graduate Texts in Mathematics*. Springer-Verlag. Part III.
28. Dan Burghelea and Nicolaas H. Kuiper. Hilbert manifolds. *Ann. of Math. (2)*, 90:379–417, 1969.
29. Henri Cartan. Théories cohomologiques. *Invent. Math.*, 35:261–271, 1976.
30. Pierre Cartier. Modules associés à un groupe formel commutatif. Courbes typiques. *C.R. Acad. Sci. Paris Sér. A-B*, 265:A129–A132, 1967.
31. Frédéric Chapoton. On a Hopf operad containing the Poisson operad. *Algebr. Geom. Topol.*, 3:1257–1273, 2003.
32. Frédéric Chapoton. Free pre-Lie algebras are free as Lie algebras, 2007. Available from http://arxiv.org/.
33. Frédéric Chapoton, Florent Hivert, Jean-Christophe Novelli, and Jean-Yves Thibon. An operational calculus for the mould operad. *Int. Math. Res. Not. IMRN*, (9):Art. ID rnn018, 22, 2008.
34. Frédéric Chapoton and Muriel Livernet. Pre-Lie algebras and the rooted trees operad. *Internat. Math. Res. Notices*, (8):395–408, 2001.
35. Frédéric Chapoton and Muriel Livernet. Relating two Hopf algebras built from an operad. *Int. Math. Res. Not. IMRN*, (24):Art. ID rnm131, 27, 2007.
36. Claude Chevalley. Review of 'Lois de groupes et analyseurs', 1956. Math. Rev. 17.
37. Joseph Chuang and Andrey Lazarev. Feynman diagrams and minimal models for operadic algebras, 2008. Available from http://arxiv.org/.
38. P. M. Cohn. *Free rings and their relations*. Academic Press, London, 1971. London Mathematical Society Monographs, No. 2.
39. Alain Connes and Dirk Kreimer. Hopf algebras, renormalization and noncommutative geometry. *Comm. Math. Phys.*, 199(1):203–242, 1998.
40. Sorin Dăscălescu, Constantin Năstăsescu, and Şerban Raianu. *Hopf algebras*, volume 235 of *Monographs and Textbooks in Pure and Applied Mathematics*. Marcel Dekker Inc., New York, 2001. An introduction.
41. Corrado de Concini and Claudio Procesi. A characteristic free approach to invariant theory. *Advances in Math.*, 21(3):330–354, 1976.
42. Michel Demazure and Pierre Gabriel. *Groupes algébriques. Tome I: Géométrie algébrique, généralités, groupes commutatifs*. Masson & Cie, Éditeur, Paris, 1970. Avec un appendice ıt Corps de classes local par Michiel Hazewinkel.
43. Richard Dipper, Stephen Doty, and Jun Hu. Brauer algebras, symplectic Schur algebras and Schur-Weyl duality. *Trans. Amer. Math. Soc.*, 360(1):189–213, 2008.

44. I. V. Dolgachev. Group scheme, 2001. in M. Hazewinkel, ed., Encyclopaedia of Mathematics. Available from http://eom.springer.de.
45. Stephen Doty. New versions of Schur-Weyl duality. In *Finite groups 2003*, pages 59–71. Walter de Gruyter, Berlin, 2004.
46. James Eells and K. D. Elworthy. On the differential topology of hilbertian manifolds. *Global Analysis, Proc. Sympos. Pure Math.*, 15:41–44, 1970.
47. James Eells and K. D. Elworthy. Open embeddings of certain Banach manifolds. *Ann. of Math. (2)*, 91:465–485, 1970.
48. Per Enflo. A counterexample to the approximation problem in Banach spaces. *Acta Math.*, 130:309–317, 1973.
49. Giovanni Felder and Alexander P. Veselov. Polynomial solutions of the Knizhnik-Zamolodchikov equations and Schur-Weyl duality. *Int. Math. Res. Not. IMRN*, (15):Art. ID rnm046, 21, 2007.
50. Yves Félix, Stephen Halperin, and Jean-Claude Thomas. *Rational homotopy theory*, volume 205 of *Graduate Texts in Mathematics*. Springer-Verlag, New York, 2001.
51. Loïc Foissy. Les algèbres de Hopf des arbres enracinés décorés. I. *Bull. Sci. Math.*, 126(3):193–239, 2002.
52. Loïc Foissy. Les algèbres de Hopf des arbres enracinés décorés. II. *Bull. Sci. Math.*, 126(4):249–288, 2002.
53. Loïc Foissy. Bidendriform bialgebras, trees, and free quasi-symmetric functions. *J. Pure Appl. Algebra*, 209(2):439–459, 2007.
54. Loïc Foissy. The infinitesimal Hopf algebra and the poset of planar forests, 2008. Available from http://arxiv.org/.
55. Loïc Foissy. Free brace algebras are free pre-Lie algebras, 2009. Available from http://arxiv.org/.
56. V. incent Franjou, Eric M. Friedlander, Teimuraz Pirashvili, and Lionel Schwartz. *Rational representations, the Steenrod algebra and functor homology*, volume 16 of *Panoramas et Synthèses [Panoramas and Syntheses]*. Société Mathématique de France, Paris, 2003.
57. Benoit Fresse. Cogroups in algebras over an operad are free algebras. *Comment. Math. Helv.*, 73(4):637–676, 1998.
58. Peter G. O. Freund and Edward Witten. Adelic string amplitudes. *Phys. Lett. B*, 199(2):191–194, 1987.
59. Ferdinand Georg Frobenius. *Gesammelte Abhandlungen. Bände I, II, III*. Herausgegeben von J.-P. Serre. Springer-Verlag, Berlin, 1968.
60. A. Fröhlich. *Formal groups*. Lecture Notes in Mathematics, No. 74. Springer-Verlag, Berlin, 1968.
61. William Fulton and Joe Harris. *Representation theory*, volume 129 of *Graduate Texts in Mathematics*. Springer-Verlag, New York, 1991. A first course, Readings in Mathematics.
62. Victor Ginzburg and Mikhail Kapranov. Koszul duality for operads. *Duke Math. J.*, 76(1):203–272, 1994.
63. Roe Goodman. Multiplicity-free spaces and Schur-Weyl-Howe duality. In *Representations of real and p-adic groups*, volume 2 of *Lect. Notes Ser. Inst. Math. Sci. Natl. Univ. Singap.*, pages 305–415. Singapore Univ. Press, Singapore, 2004.
64. Roe Goodman and Nolan R. Wallach. *Representations and invariants of the classical groups*, volume 68 of *Encyclopedia of Mathematics and its Applications*. Cambridge University Press, Cambridge, 1998.
65. J. A. Green. *Polynomial representations of* GL_n, volume 830 of *Lecture Notes in Mathematics*. Springer, Berlin, augmented edition, 2007. With an appendix on Schensted correspondence and Littelmann paths by K. Erdmann, Green and M. Schocker.
66. Luzius Grünenfelder. *Über die Struktur von Hopf-Algebren*. PhD thesis, ETH, Zürich. 85 S., 1969.
67. Alastair Hamilton and Andrey Lazarev. Characteristic classes of A_∞-algebras. *J. Homotopy Relat. Struct.*, 3(1):65–111, 2008.
68. Allen Hatcher. Vector bundles and k-theory, 2009. Available from http://www.math.cornell.edu/˜hatcher/.

69. Michiel Hazewinkel. Six chapters on Hopf Algebras. In Michiel Hazewinkel, Nadiya Gubareni, and V. V. Kirichenko, editors, *Algebras, rings and modules. Vol. 3. Series: Mathematics and Its Applications. Springer, to appear.*
70. Michiel Hazewinkel. Witt vectors. Part 1. In M. Hazewinkel, editor, *Handbook of Algebra, Volume 6, Elsevier, to appear.*
71. Michiel Hazewinkel. Constructing formal groups. I. The local one dimensional case. *J. Pure Appl. Algebra*, 9(2):131–149, 1976/77.
72. Michiel Hazewinkel. Constructing formal groups. II. The global one dimensional case. *J. Pure Appl. Algebra*, 9(2):151–161, 1976/77.
73. Michiel Hazewinkel. Constructing formal groups. III. Applications to complex cobordism and Brown-Peterson cohomology. *J. Pure Appl. Algebra*, 10(1):1–18, 1977/78.
74. Michiel Hazewinkel. *Formal groups and applications*, volume 78 of *Pure and Applied Mathematics*. Academic Press Inc. [Harcourt Brace Jovanovich Publishers], New York, 1978. Available from http://www.darenet.nl.
75. Michiel Hazewinkel. Operations in the K-theory of endomorphisms. *J. Algebra*, 84(2):285–304, 1983.
76. Michiel Hazewinkel. Cofree coalgebras and multivariable recursiveness. *J. Pure Appl. Algebra*, 183(1-3):61–103, 2003.
77. Michiel Hazewinkel. Rigidity for MPR, the Malvenuto-Poirier-Reutenauer Hopf algebra of permutations. *Honam Math. J.*, 29(4):495–509, 2007. Corrigenda and addenda: 30 (2008), no. 1, 205.
78. Michiel Hazewinkel. Towards uniqueness of MPR, the Malvenuto-Poitier-Reutenauer Hopf algebra of permutations. *Honam Math. J.*, 29(2):119–192, 2007.
79. Michiel Hazewinkel and Clyde F. Martin. Representations of the symmetric group, the specialization order, systems and Grassmann manifolds. *L'Enseignement Mathématique*, 29:53–87, 1983.
80. Howard L. Hiller. λ-rings and algebraic K-theory. *J. Pure Appl. Algebra*, 20(3):241–266, 1981.
81. Peter Hoffman. Exponential maps and λ-rings. *J. Pure Appl. Algebra*, 27(2):131–162, 1983.
82. Heinz Hopf. Über die Topologie der Gruppen-Mannigfaltigkeiten und ihre Verallgemeinerungen. *Ann. of Math. (2)*, 42:22–52, 1941.
83. Roger E. Howe. Perspectives on invariant theory: Schur duality, multiplicity-free actions and beyond. In *The Schur lectures (1992) (Tel Aviv)*, volume 8 of *Israel Math. Conf. Proc.*, pages 1–182. Bar-Ilan Univ., Ramat Gan, 1995.
84. Sze-tsen Hu. *Homotopy theory*. Pure and Applied Mathematics, Vol. VIII. Academic Press, New York, 1959.
85. Craig Huneke. Hyman Bass and ubiquity: Gorenstein rings. In *Algebra, K-theory, groups, and education (New York, 1997)*, volume 243 of *Contemp. Math.*, pages 55–78. Amer. Math. Soc., Providence, RI, 1999.
86. Dale Husemoller. *Fibre bundles*. McGraw-Hill Book Co., New York, 1966.
87. Stefan Jackowski, James McClure, and Bob Oliver. Homotopy theory of classifying spaces of compact Lie groups. In *Algebraic topology and its applications*, volume 27 of *Math. Sci. Res. Inst. Publ.*, pages 81–123. Springer, New York, 1994.
88. Daniel M. Kan. Adjoint functors. *Trans. Amer. Math. Soc.*, 87:294–329, 1958.
89. Daniel M. Kan. On monoids and their dual. *Bol. Soc. Mat. Mexicana (2)*, 3:52–61, 1958.
90. Irving Kaplansky. *Bialgebras*. Lecture Notes in Mathematics. Department of Mathematics, University of Chicago, Chicago, Ill., 1975.
91. Max Karoubi. *K-theory*. Classics in Mathematics. Springer-Verlag, Berlin, 2008. An introduction, Reprint of the 1978 edition, With a new postface by the author and a list of errata.
92. Bernhard Keller. Introduction to A-infinity algebras and modules. *Homology Homotopy Appl.*, 3(1):1–35, 2001.
93. V. K. Kharchenko. Braided version of Shirshov-Witt theorem. *J. Algebra*, 294(1):196–225, 2005.
94. A. F. Kharshiladze. Characteristic class, 2001. in M. Hazewinkel, ed., Encyclopaedia of Mathematics. Available from http://eom.springer.de.

95. Andrei Khrennikov. *p-adic valued distributions in mathematical physics*, volume 309 of *Mathematics and its Applications*. Kluwer Academic Publishers Group, Dordrecht, 1994.
96. W. H. Klink and Tuong Ton-That. Duality in representation theory. *Ulam Quarterly*, 1(1):44–52, 1992.
97. Donald Knutson. *λ-rings and the representation theory of the symmetric group*. Lecture Notes in Mathematics, Vol. 308. Springer-Verlag, Berlin, 1973.
98. Maxim Kontsevich and Yan Soibelman. Deformations of algebras over operads and the Deligne conjecture. In *Conférence Moshé Flato 1999, Vol. I (Dijon)*, volume 21 of *Math. Phys. Stud.*, pages 255–307. Kluwer Acad. Publ., Dordrecht, 2000.
99. Dirk Kreimer. On the Hopf algebra structure of perturbative quantum field theories. *Adv. Theor. Math. Phys.*, 2(2):303–334, 1998.
100. Nicolaas H. Kuiper. The homotopy type of the unitary group of Hilbert space. *Topology*, 3:19–30, 1965.
101. Alain Lascoux. *Polynômes symétriques, foncteurs de Schur et Grassmanniens*. PhD thesis, Université Paris VII, 1977.
102. Michel Lazard. Lois de groupes et analyseurs. *Ann. Sci. Ecole Norm. Sup. (3)*, 72:299–400, 1955.
103. Michel Lazard. Sur les groupes de Lie formels à un paramètre. *Bull. Soc. Math. France*, 83:251–274, 1955.
104. Michel Lazard. Lois de groupes et analyseurs. In *Séminaire Bourbaki, Vol. 3*, pages Exp. No. 109, 77–91. Soc. Math. France, Paris, 1995.
105. Andrey Lazarev. Towers of *M*U-algebras and the generalized Hopkins-Miller theorem. *Proc. London Math. Soc. (3)*, 87(2):498–522, 2003.
106. Cristian Lenart. *Combinatorial models for certain structures in Algebraic Topology and Formal Group Theory*. PhD thesis, University of Manchester, 1996.
107. Cristian Lenart. Formal group-theoretic generalizations of the necklace algebra, including a q-deformation. *J. Algebra*, 199(2):703–732, 1998.
108. Cristian Lenart. Symmetric functions, formal group laws, and Lazard's theorem. *Adv. Math.*, 134(2):219–239, 1998.
109. Jean Leray. Sur la forme des espaces topologiques et sur les points fixes des représentations. *J. Math. Pures Appl. (9)*, 24:95–167, 1945.
110. Thierry Lévy. Schur-Weyl duality and the heat kernel measure on the unitary group. *Adv. Math.*, 218(2):537–575, 2008.
111. Arunas Liulevicius. Representation rings of the symmetric groups, A Hopf algebra approach, 1975/76. Preprint Series No. 29, Aarhus University, Denmark.
112. Arunas Liulevicius. Arrows, symmetries and representation rings. *J. Pure Appl. Algebra*, 19:259–273, 1980.
113. Muriel Livernet. A rigidity theorem for pre-Lie algebras. *J. Pure Appl. Algebra*, 207(1):1–18, 2006.
114. Jean-Louis Loday and María Ronco. On the structure of cofree Hopf algebras. *J. Reine Angew. Math.*, 592:123–155, 2006.
115. J. Lukes. Sorgenfrey topology, 2001. in M. Hazewinkel, ed., Encyclopaedia of Mathematics. Available from http://eom.springer.de.
116. Saunders Mac Lane. *Categories for the working mathematician*, volume 5 of *Graduate Texts in Mathematics*. Springer-Verlag, New York, second edition, 1998.
117. I. G. Macdonald. Polynomial functors and wreath products. *J. Pure Appl. Algebra*, 18(2):173–204, 1980.
118. I. G. Macdonald. *Symmetric functions and Hall polynomials*. Oxford Mathematical Monographs. The Clarendon Press Oxford University Press, New York, second edition, 1995. With contributions by A. Zelevinsky, Oxford Science Publications.
119. Yuri I. Manin. *Mathematics as metaphor*. American Mathematical Society, Providence, RI, 2007. Selected essays of Yuri I. Manin, With a foreword by Freeman J. Dyson.
120. Martin Markl. Free homotopy algebras. *Homology Homotopy Appl.*, 7(2):123–137, 2005.

121. Martin Markl, Steve Shnider, and Jim Stasheff. *Operads in algebra, topology and physics*, volume 96 of *Mathematical Surveys and Monographs*. American Mathematical Society, Providence, RI, 2002.
122. Mitja Mastnak and Sarah Witherspoon. Bialgebra cohomology, pointed Hopf algebras, and deformations, 2008. Available from http://arxiv.org/.
123. Akira Masuoka. Classification of Semisimple Hopf Algebras. In M. Hazewinkel, editor, *Handbook of Algebra, Volume 5, Elsevier, 2008.*
124. John McCleary, editor. *Higher homotopy structures in topology and mathematical physics*, volume 227 of *Contemporary Mathematics*, Providence, RI, 1999. American Mathematical Society.
125. S. A. Merkulov. Strong homotopy algebras of a Kähler manifold. *Internat. Math. Res. Notices*, (3):153–164, 1999.
126. S. A. Merkulov. Permutahedra, hkr isomorphism and polydifferential gerstenhaber-schack complex, 2007. Available from http://arxiv.org/.
127. John W. Milnor and John C. Moore. On the structure of Hopf algebras. *Ann. of Math. (2)*, 81:211–264, 1965.
128. John W. Milnor and James D. Stasheff. *Characteristic classes*. Princeton University Press, Princeton, N.J., 1974. Annals of Mathematics Studies, No. 76.
129. Ieke Moerdijk. On the Connes-Kreimer construction of Hopf algebras. In *Homotopy methods in algebraic topology (Boulder, CO, 1999)*, volume 271 of *Contemp. Math.*, pages 311–321. Amer. Math. Soc., Providence, RI, 2001.
130. John C. Moore. Algèbres de Hopf universelles. Séminaire Henri Cartan, tome 12, *n^o* 2, (1959-1960), exposé *n^o* 10, p. 1-11. Available from http://www.numdam.org.
131. Nicole Moulis. Variétés de dimension infinie. Séminaire Bourbaki, Année 1969/1970, Exposé 378. Available from http://www.numdam.org.
132. Nicole Moulis. Sur les variétés Hilbertiennes et les fonctions non dégénérées. *Nederl. Akad. Wetensch. Proc. Ser. A 71 = Indag. Math.*, 30:497–511, 1968.
133. Jakob Nielsen. A basis for subgroups of free groups. *Math. Scand.*, 3:31–43, 1955.
134. Sergei P. Novikov. Methods of algebraic topology from the point of view of cobordism theory. *Izv. Akad. Nauk SSSR Ser. Mat.*, 31:855–951, 1967.
135. F. Patras. A Leray theorem for the generalization to operads of Hopf algebras with divided powers. *J. Algebra*, 218(2):528–542, 1999.
136. F. Patras. Leray theorem (for hopf algebras), Springer, to appear. in M. Hazewinkel, ed., Encyclopaedia of Mathematics Volume 14.
137. V. V. Prasolov. *Elements of homology theory*, volume 81 of *Graduate Studies in Mathematics*. American Mathematical Society, Providence, RI, 2007. Translated from the 2005 Russian original by Olga Sipacheva.
138. Daniel G. Quillen. On the formal group laws of unoriented and complex cobordism theory. *Bull. Amer. Math. Soc.*, 75:1293–1298, 1969.
139. Georges Racinet. *Séries génératrices non-commutatives de polyzêtas et associateurs de Drinfeld*. PhD thesis, Université de Picardie Jules Verne, Laboratoire Amiénois de Mathématique Fondamentale et Appliquée. Available from http://tel.archives-ouvertes.fr/.
140. Douglas C. Ravenel. *Complex cobordism and stable homotopy groups of spheres*, volume 121 of *Pure and Applied Mathematics*. Academic Press Inc., Orlando, FL, 1986.
141. Yu. B. Rudyak. Milnor sphere, 2001. in M. Hazewinkel, ed., Encyclopaedia of Mathematics. Available from http://eom.springer.de.
142. Otto Schreier. Die Untergruppen der freien Gruppen. *Abhandlungen Hamburg*, 5:161–183, 1927.
143. Issai Schur. *Über eine Klasse von Matrizen, die sich einer gegebenen Matrix zuordnen lassen.* PhD thesis, Diss. Berlin. 76 S , 1901.
144. Issai Schur. Über die rationalen Darstellungen der allgemeinen linearen Gruppe. *Sitzungsberichte Akad. Berlin*, 1927:58–75, 1927.
145. Heinz Georg Schuster. *Deterministic chaos*. VCH Verlagsgesellschaft mbH, Weinheim, second edition, 1988. An introduction.

146. Graeme Segal. Equivariant contractibility of the general linear group of Hilbert space. *Bull. London Math. Soc.*, 1:329–331, 1969.
147. A. I. Shirshov. Subalgebras of free Lie algebras. *Mat. Sbornik N.S.*, 33(75):441–452, 1953.
148. Frank Sottile. Schubert cell, 2001. in M. Hazewinkel, ed., Encyclopaedia of Mathematics. Available from http://eom.springer.de.
149. Lynn Arthur Steen and J. Arthur Seebach, Jr. *Counterexamples in topology*. Springer-Verlag, New York, second edition, 1978.
150. Michio Suzuki. *Group theory. I*, volume 247 of *Grundlehren der Mathematischen Wissenschaften [Fundamental Principles of Mathematical Sciences]*. Springer-Verlag, Berlin, 1982. Translated from the Japanese by the author.
151. Andrzej Szankowski. $B(\mathscr{H})$ does not have the approximation property. *Acta Math.*, 147(1-2):89–108, 1981.
152. Mitsuhiro Takeuchi. Relative Hopf modules—equivalences and freeness criteria. *J. Algebra*, 60(2):452–471, 1979.
153. Mitsuhiro Takeuchi. $\sqrt{\text{Morita}}$ theory—formal ring laws and monoidal equivalences of categories of bimodules. *J. Math. Soc. Japan*, 39(2):301–336, 1987.
154. Tai-Man Tang. Crumpled cube and solid horned sphere space fillers. *Discrete Comput. Geom.*, 31(3):421–433, 2004.
155. Daniel Tanré. *Homotopie rationnelle: modèles de Chen, Quillen, Sullivan*, volume 1025 of *Lecture Notes in Mathematics*. Springer-Verlag, Berlin, 1983.
156. Tammo tom Dieck. *Transformation groups*, volume 8 of *de Gruyter Studies in Mathematics*. Walter de Gruyter & Co., Berlin, 1987.
157. Vladimir Turaev and Hans Wenzl. Semisimple and modular categories from link invariants. *Math. Ann.*, 309(3):411–461, 1997.
158. M. Varagnolo and E. Vasserot. Schur duality in the toroidal setting. *Comm. Math. Phys.*, 182(2):469–483, 1996.
159. V. E. Voskresenskiĭ. *Algebraic groups and their birational invariants*, volume 179 of *Translations of Mathematical Monographs*. American Mathematical Society, Providence, RI, 1998. Translated from the Russian manuscript by Boris Kunyavski [Boris È. Kunyavskiĭ].
160. Stan Wagon. *The Banach-Tarski paradox*. Cambridge University Press, Cambridge, 1993. With a foreword by Jan Mycielski, Corrected reprint of the 1985 original.
161. Hermann Weyl. *The Classical Groups. Their Invariants and Representations*. Princeton University Press, Princeton, N.J., 1939.
162. Ernst Witt. Die Unterringe der freien Lieschen Ringe. *Math. Z.*, 64:195–216, 1956.
163. Steve Zelditch. Schur Weyl duality, 2006. Lecture Notes. Department of Mathematics, Johns Hopkins University.
164. Andrey V. Zelevinsky. *Representations of finite classical groups. A Hopf algebra approach*, volume 869 of *Lecture Notes in Mathematics*. Springer-Verlag, Berlin, 1981.

Chapter 6
Method of Generating Differentials

I-Chiau Huang

Abstract Enhancing the method of generating functions and the method of coefficients, various notions of differentials are brought in so that computations resulting Jacobians from changes of variables become transparent. Here we present modules of differentials for rings of formal power series and for fields of generalized power series. Local cohomology residues and logarithmic residues are reviewed with implementations on classical problems.

6.1 Introduction

This article reviews certain algebraic aspects of combinatorial phenomena, whose developments begin from searching for applications of a concrete realization [20] of Grothendieck duality and from understanding Egorychev's method of coefficients (integral representation and the computation of combinatorial sums) [9]. The reader is encouraged to look also at another review [23] discussing computations using residues, while the current article emphasizes the role played by differentials, updates some recent developments and compares related works.

For a sequence $\{a_n\}_{n\geq 0}$ of numbers with combinatorial significance, there generates a function $\sum a_n X^n$ analytic in certain region. Algebraic operations and analytic tools are available on the analytic function, from which combinatorial information can be extracted. This builds the foundation of the method of generating functions, from which the method of coefficients is developed. For these generating functions to obtain combinatorial information, manipulations include additions, multiplications, differentiations, changes of variables, substitutions of variables by functions and integrations along a path (such as contour integrals and line integrals, for instance, from 0 to 1 or from 0 to ∞), *etc.*. Of course, on the process of these ma-

I-Chiau Huang
Institute of Mathematics, Academia Sinica, Nankang, Taipei 11529, Taiwan, R.O.C.
e-mail: ichuang@math.sinica.edu.tw

I.S. Kotsireas, E.V. Zima (eds.), *Advances in Combinatorial Mathematics*,
DOI 10.1007/978-3-642-03562-3_6, © Springer-Verlag Berlin Heidelberg 2009

nipulations, one has to be aware of the convergence of the functions involved, even though these considerations have no combinatorial significance. Here we recall one of the most popular formula of analytic residues

$$\frac{1}{2\pi\sqrt{-1}} \oint \frac{\sum_{i=0}^{\infty} a_i X^i}{X^{n+1}} \mathrm{d}X = a_n$$

in the one variable case of this analytic framework. Besides thinking of this formula as a contour integral on the meromorphic function $(\sum a_i X^i)X^{-n-1}$ and then divided by $2\pi\sqrt{-1}$, it can be also considered as an operation on the meromorphic differential $(\sum a_i X^i)X^{-n-1}\mathrm{d}X$ or as an operation on the holomorphic differential $(\sum a_i X^i)\mathrm{d}X$ with n given.

Rings of (formal) power series constitute another framework. Although not equipped with all the machineries of its analytic counterpart, the elegancy and wide applicability of the algebraic framework still make it popular. As an analogue of analytic residues, there are functionals known as "coefficient of" defined on power series rings. In the one variable case, it equates the coefficient of X^n in a power series $\sum a_i X^i$ formulated as

$$[X^n]\left(\sum_{i=0}^{\infty} a_i X^i\right) = a_n.$$

In the method of coefficients, the variable X takes part in its own right in the notation

$$\mathbf{res}_X(\sum_{i=0}^{\infty} a_i X^i)X^{-n-1} = a_n.$$

We share the same view of the method of coefficients, but emphasize the role of differentials. In our approach, the formula

$$\mathrm{d}X_1 \wedge \cdots \wedge \mathrm{d}X_n = \frac{\partial(X_1, \cdots, X_n)}{\partial(Y_1, \cdots, Y_n)} \mathrm{d}Y_1 \wedge \cdots \wedge \mathrm{d}Y_n$$

is built into the framework, which naturally explains various combinatorial phenomena related to Jacobians. The concept of differentiation is used beyond the context of operators. Modules of differentials are brought to the stage so that computations involving changes of variables become transparent.

Recall that analytic residues have a cohomological interpretation. See, for instance, [16, p. 649-655]. To benefit further the algebraic convenience of the analytic theory, a realization by local cohomology for the formalism of residues can be found in [21] over arbitrary ground field. More precisely, a residue map is defined on a local cohomology module consisting of generalized fractions with an exterior product of differentials as a numerator and a system of parameters as denominators. For instance,

$$\mathbf{res}\left[\frac{(\sum_{i=0}^{\infty} a_i X^i)\mathrm{d}X}{X^{n+1}}\right] = a_n$$

with respect to a chosen variable X. Here elucidates a main difference between our approach and others. Traditionally in combinatorics, a power series ring is defined as a set, consisting of formal sums $\sum a_{i_1,\cdots,i_n} X_1^{i_1} \cdots X_n^{i_n}$, with algebraic structures. In other words, variables are already specified and included in the definition of power series rings. In our view, a power series ring is a notion free from variables; descriptions of power series depend on choices of variables; relations between two different sets of variables may contain important combinatorial information.

With respect to chosen variables, the residue map has the same effect simply as "coefficient of" functionals. However, being stable under changes of variables, the residue map turns out to be a natural tool for computations needed in many problems involving Jacobians. Along this line, Good's analytic proof [13] of MacMahon's master theorem becomes a finite process [21]; Lagrange inversion formulas are revisited [22]; pairs of inverse relations are characterized [24]. More examples can be found in [23].

As seen in [23], local cohomology residues are suitable for computations in many areas of combinatorial analysis. However, comparing to the analytic theory, there are two drawbacks. First, it does not interpret path integrals in general. Second, even though local cohomology residues work also for many cases of Laurent series, the change of variable from X to $1/X$ is not available. The second drawback can be improved by introducing a new notion of differentials [19] defined on a field consisting of "generalized power series" with exponents in a totally ordered Abelian group. A logarithmic analogue of residues to extract combinatorial information is also constructed. For instance,

$$\mathbf{res}\left[\frac{(\sum_{i=0}^{\infty} a_i X^i) X^{-n} \operatorname{dlog} X}{\log X}\right] = a_n.$$

But no cohomological interpretation is known.

For a sequence $\{a_n\}$ in a field, we specify a variable X and study the differential $(\sum a_n X^n)\mathrm{d}X$ it generates. If no changes of variables are involved, it suffices to study the generating series $\sum a_n X^n$ or the corresponding function. Three notions of differentials are available:

- holomorphic differentials and meromorphic differentials,
- finite differentials for power series rings,
- differentials for fields of generalized power series.

The analytic (holomorphic and meromorphic) differentials are quite well-known. In this article, we concentrate on algebraic (the other two notions of) differentials. For an example beyond the approach of algebraic residues, the reader is referred to the computation of a rational function summation in [11, Section 2.2], where the line integral from 0 to ∞ is used. In Sections 6.2, 6.3 and 6.4 of this article, we recall definitions and basic properties of our approach, including variables, differentials and residues. Section 6.5 consists of interpretations and viewpoints on classical problems, which are different guises of changes of variables, Schauder bases or parameters.

To enlarge the audience of this article, we include an appendix to compare our computational techniques with analytic residues [1, 27] and residues via Hochschild homology [2, 4], which have been applied to problems outside combinatorics (for instance, in effective polynomial algebra).

6.2 Variables

Some important mathematical notions usually described in terms of bases or variables (indeterminates) are in fact bases-free or variables-free. A natural setting for the idea is universal objects in categories. In this section, we revisit free Abelian groups and polynomial rings from the viewpoint of categories; power series rings are characterized as universal objects or in terms of ring-theoretic properties; fields of generalized power series are defined.

We recall some terminology in commutative algebra for reader's convenience. For more details, the reader is referred to [29]. Throughout this article, κ is a field; all the rings and algebras we consider are commutative with unity 1 and all the modules are unitary. Recall that a ring is *local*, if it has a unique maximal ideal. For instance, the formal power series ring $\kappa[[X_1,\cdots,X_n]]$ is a local ring with the maximal ideal generated by $X_1,\cdots,X_n$. An element in $\kappa[[X_1,\cdots,X_n]]$ is *invertible* if and only if it is not in the maximal ideal, equivalently, its constant term is not zero. A local ring is *complete*, if it is complete with respect to the topology given by its maximal ideal. The local ring $\kappa[[X_1,\cdots,X_n]]$ is complete. The *Krull dimension* (that is, the maximal length of chains of prime ideals) of a local ring is always finite. The complete local ring $\kappa[[X_1,\cdots,X_n]]$ has Krull dimension n.

Definition 6.1 (system of parameters). Let A be a local ring with the maximal ideal $\mathfrak{m}$ and Krull dimension n. A system of parameters consists of n elements $x_1,\cdots,x_n \in \mathfrak{m}$ generating $\mathfrak{m}$ *up to radical*. In other words, for any $z \in \mathfrak{m}$, there exist $s \in \mathbb{N}$ and $a_1,\cdots,a_n \in A$ such that

$$z^s = a_1x_1 + \cdots + a_nx_n.$$

We need the notion of systems of parameters to describe generalized fractions. Systems of parameters of a local ring always exist. A system of parameters is *regular* if the maximal ideal is generated by them. A local ring is regular if it has a regular system of parameters. The variables $X_1,\cdots,X_n$ form a regular system of parameters of the regular local ring $\kappa[[X_1,\cdots,X_n]]$.

6.2.1 Free Abelian Groups

The Abelian group $\mathbf{Z}^n$ is free of rank n. A bases-free viewpoint on free Abelian groups of rank n is provided by a category defined for a set S consisting of n elements. Let $\mathfrak{A}$ be the category whose objects are pairs (A,φ), where A is an Abelian group and $\varphi : S \to A$ is a map of sets. If (A', φ') is another object in $\mathfrak{A}$, a morphism from (A, φ) to (A', φ') is a group homomorphism $h : A \to A'$ such that $h \circ \varphi = \varphi'$. Universal objects in $\mathfrak{A}$ exist. In other words, there exists an Abelian group A, called a *free Abelian group* generated by S, together with a map $\varphi : S \to A$ satisfying the following universal property: For any object (A', φ') of $\mathfrak{A}$, there exists a unique morphism from (A, φ) to (A', φ'). The cardinality of S, called the rank of the free Abelian groups, determines the free Abelian groups up to isomorphisms. Elements in a free Abelian group A can be described by a basis, which consists of $x_1, \cdots, x_n \in A$ representing any element in A uniquely as $\sum n_i x_i$ for $n_i \in \mathbf{Z}$. Thus we regard $\mathbf{Z}^n$ as a free Abelian group of rank n *with a basis chosen*.

6.2.2 Polynomial Rings

Let $X_1, \cdots, X_n$ be indeterminates. A variables-free viewpoint on rings isomorphic to $\kappa[X_1, \cdots, X_n]$ is provided by the category described below. Let $\mathbb{N}^n$ be the product of the additive monoid of non-negative integers. Let $\mathfrak{C}$ be the category whose objects are pairs (A, φ), where A is a κ-algebra and $\varphi : \mathbb{N}^n \to A$ is a multiplicative monoid-homomorphism. If (A', φ') is another object in $\mathfrak{C}$, a morphism from (A, φ) to (A', φ') is a κ-algebra homomorphism $h : A \to A'$ such that $h \circ \varphi = \varphi'$. Universal objects in $\mathfrak{C}$ exist. For a universal object (A, φ) in $\mathfrak{C}$, we call A a *polynomial ring of n variables over κ*. Elements in such a polynomial ring A can be described using variables, that is, $X_1, \cdots, X_n \in A$ representing any element in A uniquely as a *finite* sum $\sum a_{i_1,\cdots,i_n} X_1^{i_1} \cdots X_n^{i_n}$ for $a_{i_1,\cdots,i_n} \in \kappa$. Thus we regard $\kappa[X_1, \cdots, X_n]$ as a polynomial ring of n variables over κ with *chosen variables* $X_1, \cdots, X_n$.

6.2.3 Power Series Rings

Power series again form a universal object. Let $\widehat{\mathfrak{C}}$ be the category whose objects are pairs (A, φ), where $A \supseteq \kappa$ is a complete local ring with the maximal ideal $\mathfrak{m}$ and $\varphi : \mathbb{N}^n \to \mathfrak{m}$ is a multiplicative monoid-homomorphism. If (A', φ') is another object in $\widehat{\mathfrak{C}}$, a morphism from (A, φ) to (A', φ') is a continuous (with respect to the topology given by maximal ideals) κ-algebra homomorphism $h : A \to A'$ such that $h \circ \varphi = \varphi'$. The power series ring $\kappa[[X_1, \cdots, X_n]]$ together with a homomorphism is a universal object in $\widehat{\mathfrak{C}}$. For a universal object (A, φ) in $\widehat{\mathfrak{C}}$, the ring A is called a power series ring of n-variables over κ. It can be described using variables as the case of polynomial rings. However infinite (or formal) sums are required. Thus we regard $\kappa[[X_1, \cdots, X_n]]$ as a power series ring *with chosen variables* $X_1, \cdots, X_n$. The notation

$$\kappa[[X_1, \cdots, X_n]] = \kappa[[Y_1, \cdots, Y_n]]$$

means a power series ring with two choices of sets of variables $X_1,\cdots,X_n$ and $Y_1,\cdots,Y_n$.

As a special case of Cohen's structure theorem on complete regular local rings, a power series ring A over κ can be also characterized as a complete regular local ring containing κ as a coefficient field. Assuming that the Krull dimension of A is n, we sketch a proof: Let $x_1,\cdots,x_n$ be a regular system of parameters of A. One checks directly (or apply [29, Theorem 8.4]) that the κ-algebra homomorphism $\pi : \kappa[[X_1,\cdots,X_n]] \to A$ sending X_i to x_i is surjective. A regular local ring is a domain. Hence the kernel of π is a prime ideal. The Krull dimensions of A and $\kappa[[X_1,\cdots,X_n]]$ are both n. Therefore π has to be one-to-one.

6.2.4 Fields of Generalized Power Series

Generalized power series extend power series by allowing exponents in a totally ordered Abelian group, say $\mathscr{G}$. The details of this subsection can be found in [32, Chapter 13, §2] and [19].

Definition 6.2 (generalized power series). Let $\kappa[[e^{\mathscr{G}}]]$ be the set consisting of all formal sums $\sum_{g\in\mathscr{G}} a_g e^g$, where $a_g \in \kappa$ and the support $\{g\colon a_g \neq 0\}$ is well-ordered (that is, every non-empty subset has a smallest element). An element in $\kappa[[e^{\mathscr{G}}]]$ is called a generalized power series with exponents in $\mathscr{G}$ and coefficients in κ. The scalar a_g is called the κ-coefficient of e^g in $\sum a_g e^g$.

Given $X = e^g$ with $g \in \mathscr{G}$, we use the notation

$$\log X := g.$$

For generalized power series $\sum_{g\in\mathscr{G}} a_g e^g$ and $\sum_{g\in\mathscr{G}} b_g e^g$, one shows that the multiplication

$$(\sum_{g\in\mathscr{G}} a_g e^g)(\sum_{g\in\mathscr{G}} b_g e^g) := \sum_{g\in\mathscr{G}} \left(\sum_{g_1+g_2=g} a_{g_1} b_{g_2} \right) e^g$$

is well-defined. Together with the term-wise addition

$$\sum_{g\in\mathscr{G}} a_g e^g + \sum_{g\in\mathscr{G}} b_g e^g := \sum_{g\in\mathscr{G}} (a_g + b_g) e^g,$$

$\kappa[[e^{\mathscr{G}}]]$ is a commutative ring with the unit $1 := e^0$. A generalized power series $\sum a_g e^g$ is *positive* if $a_g = 0$ for all $g \le 0$. A non-zero generalized power series can be written as $ae^g(1-\Psi)$, where $a \in \kappa$, $g \in \mathscr{G}$ and Ψ is a positive generalized power series. $a^{-1}e^{-g}(1+\Psi+\Psi^2+\cdots)$ is a well-defined generalized power series and is the inverse of $ae^g(1-\Psi)$. Hence $\kappa[[e^{\mathscr{G}}]]$ is a field.

Examples of fields of generalized power series include the field $\kappa((X))$ of Laurent series, which is isomorphic to $\kappa[[e^{\mathbf{Z}}]]$ with the usual order on $\mathbf{Z}$. The lexicographic order on $\mathbf{Z}^n$ gives rise to the field $\kappa[[e^{\mathbf{Z}^n}]]$ of iterated Laurent series.

Definition 6.3 (variable). Let $\mathscr{H}$ be a subgroup of $\mathscr{G}$ such that $\mathscr{G}/\mathscr{H}$ is free of finite rank. The elements $e^{u_1},\cdots,e^{u_n}$ are variables of $\kappa[[e^{\mathscr{G}}]]$ over $\kappa[[e^{\mathscr{H}}]]$ if and only if

$$\mathscr{G} = \mathscr{H} \oplus \mathbf{Z}u_1 \oplus \cdots \oplus \mathbf{Z}u_n.$$

As the case of power series rings, there are no canonical choices of variables! Given variables $X_1,\cdots,X_n$ of $\kappa[[e^{\mathscr{G}}]]$ over $\kappa[[e^{\mathscr{H}}]]$, every generalized power series Φ can be written uniquely as

$$\Phi = \sum_{j_1,\cdots,j_n \in \mathbf{Z}} \varphi_{j_1,\cdots,j_n} X_1^{j_1} \cdots X_n^{j_n},$$

where $\varphi_{j_1,\cdots,j_n} \in \kappa[[e^{\mathscr{H}}]]$. We call $\varphi_{0,\cdots,0}$ the *constant term* of Φ in $\kappa[[e^{\mathscr{H}}]]$. It is independent of the choice of variables.

Definition 6.4 (multiplicity). Let $X_1,\cdots,X_n$ be variables of $\kappa[[e^{\mathscr{G}}]]$ over $\kappa[[e^{\mathscr{H}}]]$ and Φ be a non-zero generalized power series with the factorization $\Phi = aY(1+\tilde{\Phi})$, where $a \in \kappa$, $\tilde{\Phi}$ is positive and $Y = e^h X_1^{i_1} \cdots X_n^{i_n}$ for some $h \in \mathscr{H}$. We call the sequence of the integers $i_1,\cdots,i_n$ the multiplicities of Φ with respect to $X_1,\cdots,X_n$.

The following definition is independent of the choice of variables.

Definition 6.5 (parameter). Non-zero generalized power series $\Phi_1,\cdots,\Phi_n$ form a system of parameters (or simply parameters) of $\kappa[[e^{\mathscr{G}}]]$ over $\kappa[[e^{\mathscr{H}}]]$ if the determinant of their multiplicities (with respect to a set of variables) is not zero in κ.

6.3 Differentials

Let A be an algebra over κ. Recall that a κ-derivation D on A into an A-module M is a κ-linear map from A to M satisfying $D(ab) = aDb + bDa$ for all $a,b \in A$. Differentials form a universal object in certain category of derivations. Reference of Subsections 6.3.1 and 6.3.2 is [26]; that of Subsection 6.3.3 is [19].

6.3.1 Kähler Differentials

There exist an A-module $\Omega_{A/\kappa}$ and a κ-derivation $\mathrm{d}: A \to \Omega_{A/\kappa}$ with the following universal property: For any A-module M and any κ-derivation $\delta: A \to M$, there exists a unique A-linear map $f: \Omega_{A/\kappa} \to M$ such that $\delta = f \circ \mathrm{d}$. We call $\Omega_{A/\kappa}$ the *module of Kähler differentials* of A over κ.

If A a polynomial ring over κ of n variables. Then $\Omega_{A/\kappa}$ is free of rank n. If $X_1,\cdots,X_n$ are variables of A over κ, then $\mathrm{d}X_1,\cdots,\mathrm{d}X_n$ form a basis for $\Omega_{A/\kappa}$; the n-th exterior power $\wedge^n \Omega_{A/\kappa}$ is free of rank 1 with $\mathrm{d}X_1 \wedge \cdots \wedge \mathrm{d}X_n$ as a basis. In such a case, the module of Kähler differentials provides a natural framework for Jacobians occurring in changes of variables. Let $Y_1, \cdots, Y_n$ be another set of variables of A. Then

$$\mathrm{d}Y_1 \wedge \cdots \wedge \mathrm{d}Y_n = \frac{\partial(X_1, \cdots, X_n)}{\partial(Y_1, \cdots, Y_n)} \mathrm{d}X_1 \wedge \cdots \wedge \mathrm{d}X_n, \tag{6.1}$$

where $\partial(Y_1, \cdots, Y_n)/\partial(X_1, \cdots, X_n)$ is the determinant of the matrix with $\partial Y_j/\partial X_i$ as the entry at the i-th row and the j-th column.

If A is a power series ring over κ, we need a notion different from Kähler differentials for Jacobians occurring in changes of variables. The reason is that the module $\Omega_{A/\kappa}$ has no easy description by variables as the case of polynomial rings. We look at an example, where K is the quotient field of $A = \kappa[[X]]$ and κ has characteristic zero. A differential basis of $\Omega_{K/\kappa}$ (that is, a subset B of K such that $\{\mathrm{d}\Phi\colon \Phi \in B\}$ forms a basis of the K-vector space $\Omega_{K/\kappa}$) is exactly a transcendence basis of K over κ [29, Theorem 26.5], whose cardinality is well-known to be infinite. Since $\Omega_{K/\kappa} \simeq \Omega_{A/\kappa} \otimes_A K$, the module $\Omega_{A/\kappa}$ is not finitely generated. Therefore $\mathrm{d}X$ does not generate $\Omega_{A/\kappa}$.

6.3.2 Finite Differentials

If M is a *finite* A-module, a κ-derivation $\mathrm{d}: A \to M$ is called *finite*. We consider the category whose objects are finite κ-derivations. Given two finite κ-derivations $\mathrm{d}: A \to M$ and $\mathrm{d}': A \to M'$, a morphism in the category is an A-linear map $M \to M'$ such that $\mathrm{d}' = f \circ \mathrm{d}$. Universal objects in this category may not exist. We use the notation $\mathrm{d}: A \to \widetilde{\Omega}_{A/\kappa}$ for a universal object in this category if it exists. We call $\widetilde{\Omega}_{A/\kappa}$ the *module of finite differentials* of A over κ.

If A a power series ring over κ with variables $X_1, \cdots, X_n$, then $\widetilde{\Omega}_{A/\kappa}$ exists and is free of rank n with a basis $\mathrm{d}X_1, \cdots, \mathrm{d}X_n$; the n-th exterior power $\wedge^n \widetilde{\Omega}_{A/\kappa}$ is free of rank 1 with $\mathrm{d}X_1 \wedge \cdots \wedge \mathrm{d}X_n$ as a basis. As the case of polynomial rings, we have formula (6.1) for two sets of variables.

6.3.3 Differentials for Generalized Power Series

Let $\mathscr{G}$ be a totally ordered Abelian group and $\mathscr{H}$ be its subgroup such that $\mathscr{G}/\mathscr{H}$ is free of rank n. We need a framework so that the formula (6.1) is available for two sets of variables.

Definition 6.6 (partial derivation). Let $X_1, \cdots, X_n$ be a set of variables of $\kappa[[e^{\mathscr{G}}]]$ over $\kappa[[e^{\mathscr{H}}]]$. The partial derivation on $\kappa[[e^{\mathscr{G}}]]$ with respect to the variables $X_1, \cdots, X_n$ is the well-defined $\kappa[[e^{\mathscr{H}}]]$-derivation

$$\frac{\partial}{\partial X_i} : \kappa[[e^{\mathscr{G}}]] \to \kappa[[e^{\mathscr{G}}]]$$

given by

$$\sum \varphi_{j_1,\cdots,j_n} X_1^{j_1} \cdots X_n^{j_n} \mapsto \sum j_i \varphi_{j_1,\cdots,j_n} X_1^{j_1} \cdots X_{i-1}^{j_{i-1}} X_i^{j_i-1} X_{i+1}^{j_{i+1}} \cdots X_n^{j_n}.$$

Definition 6.7 (chain rule). A $\kappa[[e^{\mathscr{H}}]]$-derivation D on $\kappa[[e^{\mathscr{G}}]]$ satisfies the chain rule if

$$D(\Phi) = \sum_{i=1}^{n} \frac{\partial \Phi}{\partial Y_i} D(Y_i)$$

for any $\Phi \in \kappa[[e^{\mathscr{G}}]]$ and partial derivations $\partial/\partial Y_1, \cdots, \partial/\partial Y_n$ with respect to any set of variables $Y_1, \cdots, Y_n$.

To define differentials, we consider a full subcategory of the category of $\kappa[[e^{\mathscr{H}}]]$-derivations on $\kappa[[e^{\mathscr{G}}]]$, whose objects are those derivations (for instance, partial derivations) satisfying the chain rule.

Definition 6.8 (differential). A differential of $\mathscr{G}$ over $\mathscr{H}$ is an element of a $\kappa[[e^{\mathscr{G}}]]$-vector space $\Omega_{\mathscr{G}/\mathscr{H}}$, which comes from a universal object $\mathrm{d}\colon \kappa[[e^{\mathscr{G}}]] \to \Omega_{\mathscr{G}/\mathscr{H}}$ of the subcategory described above.

A universal object $\Omega_{\mathscr{G}/\mathscr{H}}$ exists with $\mathrm{d}X_1, \cdots, \mathrm{d}X_n$ as a basis for any set of variables $X_1, \cdots, X_n$. Differentials of $\mathscr{G}$ over $\mathscr{H}$ are different from Kähler differentials of $\kappa[[e^{\mathscr{G}}]]$ over $\kappa[[e^{\mathscr{H}}]]$. For example, $\Omega_{\mathbf{Z}/0}$ is a one-dimensional $\kappa[[e^{\mathbf{Z}}]]$-vector space with the usual order on $\mathbf{Z}$. As discussed in Subsection 6.3.1, the dimension of the $\kappa[[e^{\mathbf{Z}}]]$-vector space $\Omega_{\kappa[[e^{\mathbf{Z}}]]/\kappa}$ is infinite if the characteristic of κ is zero. This argument shows that there do exist $\kappa[[e^{\mathscr{H}}]]$-derivations not satisfying the chain rule.

Given a non-zero generalized power series Υ, we use the notation

$$\mathrm{dlog}\,\Upsilon := \frac{\mathrm{d}\Upsilon}{\Upsilon}.$$

For parameters $\Phi_1, \cdots, \Phi_n$, the differentials $\mathrm{dlog}\,\Phi_1, \cdots, \mathrm{dlog}\,\Phi_n$ form a basis for the $\kappa[[e^{\mathscr{G}}]]$-vector space $\Omega_{\mathscr{G}/\mathscr{H}}$; the exterior product

$$\mathrm{dlog}\,\boldsymbol{\Phi} := \mathrm{dlog}\,\Phi_1 \wedge \cdots \wedge \mathrm{dlog}\,\Phi_n$$

forms a basis of the $\kappa[[e^{\mathscr{G}}]]$-vector space $\wedge^n \Omega_{\mathscr{G}/\mathscr{H}}$. We can extend the notion of partial derivations from the context of variables to parameters: For any non-zero generalized power series Υ, we define $\partial \log \Upsilon / \partial \log \Phi_i$ to be the unique generalized power series such that

$$\mathrm{dlog}\,\Upsilon = \frac{\partial \log \Upsilon}{\partial \log \Phi_1} \mathrm{dlog}\,\Phi_1 + \cdots + \frac{\partial \log \Upsilon}{\partial \log \Phi_n} \mathrm{dlog}\,\Phi_n.$$

For variables $X_1, \cdots, X_n$, we have the formula

$$\frac{\partial \log \Upsilon}{\partial \log X_i} = \frac{X_i}{\Upsilon} \frac{\partial \Upsilon}{\partial X_i}.$$

Let $\Psi_1, \cdots, \Psi_n$ be another system of parameters. Then

$$\mathrm{dlog}\,\boldsymbol{\Phi} = \frac{\partial \log \boldsymbol{\Phi}}{\partial \log \boldsymbol{\Psi}}\,\mathrm{dlog}\,\boldsymbol{\Psi} := \frac{\partial \log(\Phi_1, \cdots, \Phi_n)}{\partial \log(\Psi_1, \cdots, \Psi_n)}\,\mathrm{dlog}\,\boldsymbol{\Psi},$$

where $\partial \log(\Phi_1, \cdots, \Phi_n)/\partial \log(\Psi_1, \cdots, \Psi_n)$ is the determinant of the matrix with $\partial \log \Phi_i / \partial \log \Psi_j$ as the entry at the i-th row and the j-th column.

6.4 Residues

In this section, we define residues for generalized fractions in the contexts of rings of power series and fields of generalized power series. In both contexts, the numerator of a generalized fraction is an exterior product of differentials. Given variables $X_1, \cdots, X_n$, we use the shorthand

$$\mathrm{d}X := \mathrm{d}X_1 \cdots \mathrm{d}X_n := \mathrm{d}X_1 \wedge \cdots \wedge \mathrm{d}X_n.$$

6.4.1 Local Cohomology Residues

References of this subsection are [20, 23]. Let A be a power series ring over κ of n variables. The elements of the n-th local cohomology module $\mathrm{H}^n_{\mathfrak{m}}(\wedge^n \widetilde{\Omega}_{A/\kappa})$ supported at the maximal ideal $\mathfrak{m}$ of A are called *generalized fractions*. They are of the form

$$\begin{bmatrix} \omega \\ \phi_1, \cdots, \phi_n \end{bmatrix},$$

where the numerator $\omega \in \wedge^n \widetilde{\Omega}_{A/\kappa}$ and the denominators $\phi_1, \cdots, \phi_n$ form a system of parameters of A (see Definition 6.1). Generalized fractions enjoy the following properties:

Linearity Law. For $\omega_1, \omega_2 \in \wedge^n \widetilde{\Omega}_{A/\kappa}$, $a_1, a_2 \in A$ and a system of parameters $\phi_1, \cdots, \phi_n$,

$$\begin{bmatrix} a_1\omega_1 + a_2\omega_2 \\ \phi_1, \cdots, \phi_n \end{bmatrix} = a_1 \begin{bmatrix} \omega_1 \\ \phi_1, \cdots, \phi_n \end{bmatrix} + a_2 \begin{bmatrix} \omega_2 \\ \phi_1, \cdots, \phi_n \end{bmatrix}.$$

Transformation Law. For $\omega \in \wedge^n \widetilde{\Omega}_{A/\kappa}$ and two systems of parameters $\phi_1, \cdots, \phi_n$ and $\phi'_1, \cdots, \phi'_n$,

$$\begin{bmatrix} \omega \\ \phi_1, \cdots, \phi_n \end{bmatrix} = \begin{bmatrix} \det(r_{ij})\omega \\ \phi'_1, \cdots, \phi'_n \end{bmatrix}$$

if $\phi'_i = \sum_{j=1}^n r_{ij}\phi_j$ for $i = 1, \cdots, n$.

Vanishing Law. For $\omega \in \wedge^n \widetilde{\Omega}_{A/\kappa}$ and a system of parameters $\phi_1, \cdots, \phi_n$,

$$\begin{bmatrix} \omega \\ \phi_1, \cdots, \phi_n \end{bmatrix} = 0$$

if and only if $\omega \in (\phi_1, \cdots, \phi_n) \wedge^n \widetilde{\Omega}_{A/\kappa}$.

Recall that, if $\phi_1, \cdots, \phi_n$ is a system of parameters, so is $\phi_1^{i_1}, \cdots, \phi_n^{i_n}$ for any positive integers $i_1, \cdots, i_n$. For convenience, we use the convention

$$\begin{bmatrix} \omega \\ \phi_1^{i_1}, \cdots, \phi_n^{i_n} \end{bmatrix} = 0$$

if some of the exponents $i_1, \cdots, i_n$ are negative or zero.

Assume that $A = \kappa[[X_1, \cdots, X_n]]$. Using the linearity law and the vanishing law, every element in $\mathrm{H}^n_{\mathfrak{m}}(\wedge^n \tilde{\Omega}_{A/\kappa})$ can be written uniquely as a κ-linear (finite) combination of the generalized fractions of the form

$$\begin{bmatrix} \mathrm{d}X_1 \cdots \mathrm{d}X_n \\ X_1^{i_1}, \cdots, X_n^{i_n} \end{bmatrix},$$

where $i_1, \cdots, i_n$ are positive integers.

Definition 6.9 (residue). Assume that $A = \kappa[[X_1, \cdots, X_n]]$. The residue map

$$\mathbf{res}_{X_1, \cdots, X_n} : \mathrm{H}^n_{\mathfrak{m}}(\wedge^n \tilde{\Omega}_{A/\kappa}) \to \kappa$$

is defined to be the κ-linear map satisfying

$$\mathbf{res}_{X_1, \cdots, X_n} \begin{bmatrix} \mathrm{d}X_1 \cdots \mathrm{d}X_n \\ X_1^{i_1}, \cdots, X_n^{i_n} \end{bmatrix} = \begin{cases} 1, & \text{if } i_1 = \cdots = i_n = 1; \\ 0, & \text{otherwise.} \end{cases}$$

We may write the residue map without the subscript, since it is independent of the choices of variables:

Invariance Law. If $\kappa[[X_1, \cdots, X_n]] = \kappa[[Y_1, \cdots, Y_n]]$, then $\mathbf{res}_{X_1, \cdots, X_n} = \mathbf{res}_{Y_1, \cdots, Y_n}$.

Using modules of zero dimensional support, the notion of residues can be extended to power series rings over a complete local ring [20, Chapter 5]. In such a general context, residue maps are transitive. In practice, the following formula is useful.

Transitivity Law. Given $f_{i_1, \cdots, i_m} \in \kappa[[X_1, \cdots, X_n]]$,

$$\mathbf{res} \begin{bmatrix} (\sum f_{i_1, \cdots, i_m} Y_1^{i_1} \cdots Y_m^{i_m}) \mathrm{d}X_1 \cdots \mathrm{d}X_n \mathrm{d}Y_1 \cdots \mathrm{d}Y_m \\ X_1^{s_1+1}, \cdots, X_n^{s_n+1}, Y_1^{t_1+1}, \cdots, Y_m^{t_m+1} \end{bmatrix} = \mathbf{res} \begin{bmatrix} f_{t_1, \cdots, t_m} \mathrm{d}X_1 \cdots \mathrm{d}X_n \\ X_1^{s_1+1}, \cdots, X_n^{s_n+1} \end{bmatrix}.$$

6.4.2 Logarithmic Residues

Reference of this subsection is [19], where omitted proofs can be found. In the set

$$\{(\alpha, \Phi_1, \cdots, \Phi_n) \in \wedge^n \Omega_{\mathscr{G}/\mathscr{H}} \times \kappa[[e^{\mathscr{G}}]]^n : \ \Phi_1, \cdots, \Phi_n \text{ are parameters}\},$$

we define an equivalence relation:

$$(\alpha, \Phi_1, \cdots, \Phi_n) \sim (\beta, \Psi_1, \cdots, \Psi_n) \Longleftrightarrow \frac{\beta}{\det t_{ij}} = \det u_{ij} \frac{\alpha}{\det s_{ij}},$$

where s_{ij} (resp. t_{ij}) are multiplicities of Φ_i (resp. Ψ_i) with respect to a set of variables $X_1, \cdots, X_n$ (resp. $Y_1, \cdots, Y_n$) and $Y_i = e^{h_i} X_1^{u_{i1}} \cdots X_n^{u_{in}}$. The equivalence relation is independent of the choices of variables.

Definition 6.10 (generalized fraction). A generalized fraction

$$\begin{bmatrix} \alpha \\ \log \boldsymbol{\Phi} \end{bmatrix} := \begin{bmatrix} \alpha \\ \log \Phi_1, \cdots, \log \Phi_n \end{bmatrix}$$

is the equivalence class containing $(\alpha, \Phi_1, \cdots, \Phi_n)$. We call α the numerator of the generalized fraction. The set of generalized fractions is denoted by $\mathrm{H}(\wedge^n \Omega_{\mathscr{G}/\mathscr{H}})$.

Generalized fractions enjoy the following properties:

Linearity Law. For $\omega_1, \omega_2 \in \wedge^n \Omega_{\mathscr{G}/\mathscr{H}}$, $\Psi_1, \Psi_2 \in \kappa[[e^{\mathscr{G}}]]$ and parameters $\Phi_1, \cdots, \Phi_n$,

$$\begin{bmatrix} \Psi_1 \omega_1 + \Psi_2 \omega_2 \\ \log \boldsymbol{\Phi} \end{bmatrix} = \Psi_1 \begin{bmatrix} \omega_1 \\ \log \boldsymbol{\Phi} \end{bmatrix} + \Psi_2 \begin{bmatrix} \omega_2 \\ \log \boldsymbol{\Phi} \end{bmatrix}.$$

Transformation Law. For $\omega \in \wedge^n \Omega_{\mathscr{G}/\mathscr{H}}$ and two systems of parameters $\Phi_1, \cdots, \Phi_n$ and $\Phi'_1, \cdots, \Phi'_n$,

$$\begin{bmatrix} \omega \\ \log \boldsymbol{\Phi} \end{bmatrix} = \begin{bmatrix} \det(r_{ij}) \omega \\ \log \boldsymbol{\Phi}' \end{bmatrix}$$

if $\Phi'_i = \prod_{j=1}^n \Phi_j^{r_{ij}}$ for $i = 1, \cdots, n$.

Definition 6.11 (residue). Given variables $X_1, \cdots, X_n$ of $\kappa[[e^{\mathscr{G}}]]$ over $\kappa[[e^{\mathscr{H}}]]$, we define the residue map

$$\mathbf{res}_{X_1, \cdots, X_n} : \ \mathrm{H}(\wedge^n \Omega_{\mathscr{G}/\mathscr{H}}) \to \kappa[[e^{\mathscr{H}}]]$$

by

$$\mathbf{res}_{X_1, \cdots, X_n} \begin{bmatrix} \Phi \, \mathrm{d}\log \mathbf{X} \\ \log \mathbf{X} \end{bmatrix} = \text{ the constant term of } \Phi \text{ in } \kappa[[e^{\mathscr{H}}]],$$

where $\Phi \in \kappa[[e^{\mathscr{G}}]]$.

We may write the residue map without the subscript, since it is independent of the choices of variables:

Invariance Law. $\mathbf{res}_{X_1, \cdots, X_n} = \mathbf{res}_{Y_1, \cdots, Y_n}$ for any two sets of variables $X_1, \cdots, X_n$ and $Y_1, \cdots, Y_n$ of $\kappa[[e^{\mathscr{G}}]]$ over $\kappa[[e^{\mathscr{H}}]]$.

Let $\Phi_1, \cdots, \Phi_n \in \kappa[[e^{\mathscr{G}}]]$ and $\psi_{i_1,\cdots,i_n} \in \kappa[[e^{\mathscr{H}}]]$, where the indices $i_\ell \in \mathbb{Z}$. We assume that there are only finitely many $\psi_{i_1,\cdots,i_n}\Phi_1^{i_1}\cdots\Phi_n^{i_n}$ whose support contains any fixed $g \in \mathscr{G}$. Under such an assumption, we can define a set $\operatorname{supp}\Psi(\boldsymbol{\Phi})$ consisting of those $g \in \mathscr{G}$ such that the sum of the κ-coefficients of e^g in $\psi_{i_1,\cdots,i_n}\Phi_1^{i_1}\cdots\Phi_n^{i_n}$ is nonzero. We assume furthermore that $\operatorname{supp}\Psi(\boldsymbol{\Phi})$ is well-ordered. Under these assumptions, we can define an element

$$\Psi(\boldsymbol{\Phi}) := \sum \psi_{i_1,\cdots,i_n}\Phi_1^{i_1}\cdots\Phi_n^{i_n} \in \kappa[[e^{\mathscr{G}}]],$$

so that the κ-coefficient of e^g in $\Psi(\boldsymbol{\Phi})$ is the sum of those in $\psi_{i_1,\cdots,i_n}\Phi_1^{i_1}\cdots\Phi_n^{i_n}$. We say that $\Psi(\boldsymbol{\Phi})$ is represented by $\Phi_1, \cdots, \Phi_n$ with $\psi_{i_1,\cdots,i_n}$ as the $\kappa[[e^{\mathscr{H}}]]$-coefficient of $\Phi_1^{i_1}\cdots\Phi_n^{i_n}$. The residue map also respects representations of a generalized fraction by parameters:

Jacobi Law. Given a representation $\Psi(\boldsymbol{\Phi}) = \sum \psi_{i_1,\cdots,i_n}\Phi_1^{i_1}\cdots\Phi_n^{i_n} \in \kappa[[e^{\mathscr{G}}]]$ by parameters $\Phi_1, \cdots, \Phi_n$ with $\psi_{i_1,\cdots,i_n} \in \kappa[[e^{\mathscr{H}}]]$,

$$\mathbf{res}\begin{bmatrix}\Psi(\boldsymbol{\Phi})\,\mathrm{dlog}\,\boldsymbol{\Phi}\\ \log\boldsymbol{\Phi}\end{bmatrix} = \psi_{0,\cdots,0}.$$

Jacobi law is essentially a generalization of a formula of Jacobi [25] (see [12, §5] for a historical account), which deals with elements $\Phi_1, \cdots, \Phi_n$ in a field of iterated Laurent series over κ with variables $X_1, \cdots, X_n$. Assume that, for each i, there exist $s_{ij} \in \mathbb{Z}$ such that $\Phi_i/(X_1^{s_{i1}}\cdots X_n^{s_{in}})$ is a power series with nonzero constant term. Given $a_{i_1,\cdots,i_n} \in \kappa$ such that $\Upsilon(\boldsymbol{\Phi}) := \sum a_{i_1,\cdots,i_n}\Phi_1^{i_1}\cdots\Phi_n^{i_n}$ represents an iterated Laurent series, Jacobi's formula asserts that the coefficient of $X_1^{-1}\cdots X_n^{-1}$ in $\Upsilon(\boldsymbol{\Phi})|\partial\boldsymbol{\Phi}/\partial\mathbf{X}|$ is $a_{-1,\cdots,-1}\det s_{ij}$. For the case that $\det s_{ij}$ is nonzero in κ (that is, $\Phi_1, \cdots, \Phi_n$ are parameters), we can write

$$\begin{bmatrix}\boldsymbol{\Phi}\Upsilon(\boldsymbol{\Phi})\,\mathrm{dlog}\,\boldsymbol{\Phi}\\ \log\boldsymbol{\Phi}\end{bmatrix} = \begin{bmatrix}\frac{\mathbf{X}}{\det s_{ij}}\Upsilon(\boldsymbol{\Phi})\left|\frac{\partial\boldsymbol{\Phi}}{\partial\mathbf{X}}\right|\mathrm{dlog}\,\mathbf{X}\\ \log\mathbf{X}\end{bmatrix}.$$

Apply the residue map, together with Jacobi Law, we recover the required formula

$$\mathbf{res}\begin{bmatrix}\frac{\mathbf{X}}{\det s_{ij}}\Upsilon(\boldsymbol{\Phi})\left|\frac{\partial\boldsymbol{\Phi}}{\partial\mathbf{X}}\right|\mathrm{dlog}\,\mathbf{X}\\ \log\mathbf{X}\end{bmatrix} = a_{-1,\cdots,-1}.$$

If $\det s_{ij}$ vanishes in κ, the formulation of our Jacobi law is no longer available. However, the proof of Jacobi law in [19] still gives rise to Jacobi's formula.

6.5 Implementations

In this section, we explain our viewpoints on inverting combinatorial sums, compositional inverses and Lagrange inversions. MacMahon's master theorem and Dyson's conjecture are revisited. An example is given to exhibit the irrelevance of analytic constraints on combinatorial problems.

6.5.1 Inverting Combinatorial Sums

Given a combinatorial sum

$$b_n = \sum_{k=0}^{n} c_{nk} a_k, \tag{6.2}$$

where $a_i, b_i, c_{ji} \in \kappa$, Riordan [33] considers the general problem of inverting it, that is, finding $d_{ij} \in \kappa$ such that

$$a_n = \sum_{k=0}^{n} d_{nk} b_k.$$

We consider the case that $c_{ii} \neq 0$ for all i and assume that $c_{ij} = 0$ for $i < j$. A solution to the problem of inverting the combinatorial sum (6.2) is given by the lower triangular matrix (d_{ij}) which is the inverse of (c_{ij}).

The problem can be also stated using Schauder bases [24]: Let (f_j) be a strictly monotone Schauder basis of the power series ring $\kappa[[X]]$ given by

$$f_j = c_{0j} + c_{1j}X + c_{2j}X^2 + c_{3j}X^3 + \cdots$$

(that is, $c_{ii} \neq 0$ for all i and $c_{ij} = 0$ for $i < j$). Then the ordinary Schauder basis (X^j) can be written uniquely as

$$X^j = d_{0j}f_0 + d_{1j}f_1 + d_{2j}f_2 + d_{3j}f_3 + \cdots,$$

which provides a solution to (6.2). See the proof of [24, Theorem 2.1]. For some special forms of f_i, the scalars d_{ij} can be computed using local cohomology residues.

For instance, if f_j is of the form $f_j = \eta Y^j$ for an invertible power series η and a variable Y, then the scalars d_{ij} can be computed as follows:

$$d_{ij} = \mathbf{res}\left[\frac{X^j \mathrm{d}Y}{\eta Y^{i+1}}\right] = \mathbf{res}\left[\frac{(\frac{X}{Y})^{i+1}\frac{\mathrm{d}Y}{\mathrm{d}X}\mathrm{d}X}{\eta X^{i-j+1}}\right].$$

Such a special case of Schauder bases corresponds to a proper Riordan array (η, Y), *cf.* [31, Theorem 4]. For the schauder bases $(X^i e^{(p+i)X})$, $(X^i(X/(e^X-1))^{p+i})$ (assuming that the characteristic of κ is zero) and $(X^i/(1-X)^{p+i+1})$, the computations are carried out in [24, Section 3]. See [31, Section 2] for more computations of inversions listed in Riordan's book [33].

Now we consider the case called implicit inversions in [31]. Let

$$c_{ij} = \mathbf{res}\left[\frac{\eta\phi^i Z^j \mathrm{d}Y}{Y^{i+1}}\right],$$

where η and ϕ are invertible elements in $\kappa[[Z]] = \kappa[[Y]]$. In terms of the variable $X := Y/\phi$,

$$c_{ij} = \mathbf{res}\left[\frac{\eta Z^j \frac{\mathrm{d}Y}{\mathrm{d}X}\mathrm{d}X}{\phi X^{i+1}}\right].$$

Hence $f_j := \sum_i c_{ij}X^i = \eta\phi^{-1}(\mathrm{d}Y/\mathrm{d}X)Z^j$ is of the form in the previous paragraph. From the relation

$$\frac{\mathrm{d}X}{\mathrm{d}Y} = \phi^{-1} - Y\phi^{-2}\frac{\mathrm{d}\phi}{\mathrm{d}Y}, \tag{6.3}$$

we recover the formula

$$d_{ij} = \mathbf{res}\left[\frac{X^j\phi\frac{\mathrm{d}X}{\mathrm{d}Y}\mathrm{d}Z}{\eta Z^{i+1}}\right] = \mathbf{res}\left[\frac{(\phi - Y\frac{d\phi}{\mathrm{d}Y})(\frac{Y}{Z})^{i+1}\frac{\mathrm{d}Z}{\mathrm{d}Y}\mathrm{d}Y}{\eta\phi^{j+1}Y^{i-j+1}}\right]$$

in [31, Theorem 6]. The interplay of the variables X and Y is the theme of Lagrange inversions, which is built into our framework. Further discussions will be given in the next subsection.

Another aspect of inverting combinatorial sums is the notion of inverse relations. Recall that an inverse relation is a pair of identities of the form

$$\begin{cases} b_n = \sum_{k=0}^{n} c_{nk}a_k, \\ a_n = \sum_{k=0}^{n} d_{nk}b_k, \end{cases}$$

where $a_i, b_i, c_{ji}, d_{ji} \in \kappa$. The first complete characterization of inverse relations was given in [7, 8, 9]. See also [10]. Inverse relations with the condition $c_{ij} = d_{ij} = 0$ for $i < j$ and orthogonal relation

$$\sum_{k=0}^{\infty} c_{mk}d_{kn} = \delta_{mn}$$

can be also characterized by strictly monotone Schauder bases [24, Theorem 2.1].

6.5.2 Compositional Inverses and Lagrange Inversions

Consider a power series ring $\kappa[[X]]$. Recall that the *compositional inverse* (with respect to the variable X) of a given power series f is a power series $\bar{f}$ such that $f(\bar{f}(X)) = \bar{f}(f(X)) = X$. One can show that f has a compositional inverse if and only if

$$\mathbf{res}\begin{bmatrix} f\mathrm{d}X \\ X \end{bmatrix} = 0 \neq \mathbf{res}\begin{bmatrix} \mathrm{d}f \\ X \end{bmatrix}.$$

The notion of compositional inverses depends on a choice of a variable.

Let $X_1, \cdots, X_n$ and $Y_1, \cdots, Y_n$ be two sets of variables of a power series ring with the relation

$$Y_j = \sum c^{(j)}_{i_1,\cdots,i_n} X_1^{i_1} \cdots X_n^{i_n},$$

where $c^{(j)}_{i_1,\cdots,i_n} \in \kappa$. For a power series f represented as

$$f = \sum a_{i_1,\cdots,i_n} X_1^{i_1} \cdots X_n^{i_n}$$

with $a_{i_1,\cdots,i_n} \in \kappa$, *Lagrange inversion* seeks formulas of $b_{i_1,\cdots,i_n} \in \kappa$ in terms of $c^{(j)}_{i_1,\cdots,i_n}$ and $a_{i_1,\cdots,i_n}$ for a new representation

$$f = \sum b_{i_1,\cdots,i_n} Y_1^{i_1} \cdots Y_n^{i_n}.$$

From our viewpoint, computing compositional inverses and finding Lagrange inversion formulas in the one variable case are essentially the same thing. Note that a power series $Y \in \kappa[[X]]$ has a compositional inverse if and only if Y is a variable. The compositional inverse $\bar{f}$ of Y simply represents X in terms of Y, that is, $X = \bar{f}(Y)$. To obtain the compositional inverse, the coefficient $\bar{a}_n$ of Y^n in $\bar{f}$ can be computed by

$$\bar{a}_n = \mathbf{res}\begin{bmatrix} X\mathrm{d}Y \\ Y^{n+1} \end{bmatrix}.$$

Keeping track of changes of variables, Lagrange inversions are naturally formulated. We revisit a version of Lagrange inversion formula for the case $Y = X\phi$, where ϕ is an invertible power series. It is stated in [30] that

$$[X^n]Y^k = \frac{k}{n}[X^{n-k}]\phi^n(X). \tag{6.4}$$

Our view on (6.4) starts with the power series ring $\kappa[[X]] = \kappa[[Y]]$, where X and Y satisfying a relation are not treated as dummy variables. The left-hand side of (6.4) is

$$\mathbf{res}\begin{bmatrix} Y^k\mathrm{d}X \\ X^{n+1} \end{bmatrix} = \frac{1}{n}\mathbf{res}\begin{bmatrix} \mathrm{d}Y^k \\ X^n \end{bmatrix}.$$

In terms of Y,

$$\mathbf{res}\begin{bmatrix} \mathrm{d}Y^k \\ X^n \end{bmatrix} = k\,\mathbf{res}\begin{bmatrix} Y^{k-1}\mathrm{d}Y \\ X^n \end{bmatrix} = k\,\mathbf{res}\begin{bmatrix} \phi^n\mathrm{d}Y \\ Y^{n-k+1} \end{bmatrix}.$$

In our view, formula (6.4) stated as

$$[X^n]Y^k = \frac{k}{n}[Y^{n-k}]\phi^n(Y)$$

sheds light to its nature.

The Lagrange inversion formula in the so-called diagonalization form [15, p. 17] deals with power series $\eta, \phi \in \kappa[[Y]]$ with ϕ invertible. In terms of the new variable $X := Y/\phi$, the coefficient a_n of Y^n in $\eta\phi^n$ can be written as

$$a_n = \mathbf{res}\left[\frac{\eta\phi^n \mathrm{d}Y}{Y^{n+1}}\right] = \mathbf{res}\left[\frac{\eta\phi^{-1}\mathrm{d}Y}{X^{n+1}}\right] = \mathbf{res}\left[\frac{\eta \mathrm{d}X}{(1-X\frac{d\phi}{\mathrm{d}Y})X^{n+1}}\right],$$

where the last equality is by relation (6.3).

More Lagrange inversion formulas can be computed using our method. See [22] for details.

6.5.3 MacMahon's Master Theorem

Coefficients in two power series of n variables sometimes are naturally related through a power series of $2n$ variables. For an example, we interpret Egorychev's proof [9] of MacMahon's master theorem using the ring $\kappa[[z_1,\cdots,z_n,t_1,\cdots,t_n]]$, through which subrings $\kappa[[z_1,\cdots,z_n]]$ and $\kappa[[t_1,\cdots,t_n]]$ are related. We remark that the idea of duplication of variables has been also developed for different purposes [2, 4]. See Proposition 6.3 and the proof of Proposition 6.2 in the Appendix.

MacMahon's Master Theorem. Given positive integers $m_1,\cdots,m_n$ and $a_{ij} \in \kappa$, we denote $f_i = \sum_{j=1}^n a_{ij}z_j$ and $\Delta = \det(\delta_{ij} - a_{ij}t_i)$, where δ_{ij} is the Kronecker symbol. Then

$$\mathbf{res}\left[\begin{matrix} f_1^{m_1}\cdots f_n^{m_n}\mathrm{d}z \\ z_1^{m_1+1},\cdots,z_n^{m_n+1}\end{matrix}\right] = \mathbf{res}\left[\begin{matrix} \Delta^{-1}\mathrm{d}t \\ t_1^{m_1+1},\cdots,t_n^{m_n+1}\end{matrix}\right].$$

Proof. The missing link between the two residues in the theorem is provided by $y_i := z_i - t_if_i$. Note that $\kappa[[y_1,\cdots,y_n,t_1,\cdots,t_n]] = \kappa[[z_1,\cdots,z_n,t_1,\cdots,t_n]]$ and $\mathrm{d}y\mathrm{d}t = \Delta\mathrm{d}z\mathrm{d}t$. The theorem follows from

$$\mathbf{res}\left[\begin{matrix} f_1^{m_1}\cdots f_n^{m_n}\mathrm{d}z \\ z_1^{m_1+1},\cdots,z_n^{m_n+1}\end{matrix}\right] = \mathbf{res}\left[\begin{matrix} \mathrm{d}z\mathrm{d}t \\ y_1,\cdots,y_n,t_1^{m_1+1},\cdots,t_n^{m_n+1}\end{matrix}\right],$$

which is a special case of the following formula from [9, Theorem 5.4.1] for $k=1$, $\varphi_i = z_i$, $f_i = \sum_{j=1}^n a_{ij}z_j$ and $\psi = z_1^{m_1-1}\cdots z_n^{m_n-1}$. □

Formula. Given $\psi, f_1,\cdots,f_n \in \kappa[[z_1,\cdots,z_n]]$, a system of parameters $\varphi_1,\cdots,\varphi_n$ of $\kappa[[z_1,\cdots,z_n]]$ and positive integers $m_1,\cdots,m_n$,

$$\mathbf{res}\left[\begin{matrix} (f_1^{m_1}\cdots f_n^{m_n})^k\psi\mathrm{d}z \\ \varphi_1^{(k+1)m_1},\cdots,\varphi_n^{(k+1)m_n}\end{matrix}\right]$$

$$= \mathbf{res}\left[\begin{matrix} \psi\mathrm{d}z\mathrm{d}t \\ \varphi_1^{m_1}-t_1f_1^{m_1},\cdots,\varphi_n^{m_n}-t_nf_n^{m_n},t_1^{k+1},\cdots,t_n^{k+1}\end{matrix}\right] =$$

$$= \mathbf{res}\begin{bmatrix} (\varphi_1\cdots\varphi_n)^k\psi \mathrm{d}z\mathrm{d}t \\ \varphi_1^{m_1}(\varphi_1^k - t_1 f_1^k),\cdots,\varphi_n^{m_n}(\varphi_n^k - t_n f_n^k), t_1^{m_1+1},\cdots,t_n^{m_n+1}\end{bmatrix}$$

Proof. For the first identity, we observe that

$$\begin{aligned} &\begin{bmatrix} \psi \mathrm{d}z\mathrm{d}t_1 \\ \varphi_1^{m_1} - t_1 f_1^{m_1}, \varphi_2^{k+1},\cdots,\varphi_n^{k+1}, t_1^{k+1}\end{bmatrix} \\ =&\begin{bmatrix} \left(\sum_{i=0}^k (\varphi_1^{m_1})^i (t_1 f_1^{m_1})^{k-i}\right)\psi \mathrm{d}z\mathrm{d}t_1 \\ \varphi_1^{(k+1)m_1} - t_1^{k+1} f_1^{(k+1)m_1}, \varphi_2^{k+1},\cdots,\varphi_n^{k+1}, t_1^{k+1}\end{bmatrix} \\ =&\begin{bmatrix} \left(\sum_{i=0}^k (\varphi_1^{m_1})^i (t_1 f_1^{m_1})^{k-i}\right)\psi \mathrm{d}z\mathrm{d}t_1 \\ \varphi_1^{(k+1)m_1}, \varphi_2^{k+1},\cdots,\varphi_n^{k+1}, t_1^{k+1}\end{bmatrix}. \end{aligned}$$

Applying the residue map, we get

$$\mathbf{res}\begin{bmatrix} \psi \mathrm{d}z\mathrm{d}t_1 \\ \varphi_1^{m_1} - t_1 f_1^{m_1}, \varphi_2^{k+1},\cdots,\varphi_n^{k+1}, t_1^{k+1}\end{bmatrix} = \mathbf{res}\begin{bmatrix} (f_1^{m_1})^k\psi \mathrm{d}z \\ \varphi_1^{(k+1)m_1}, \varphi_2^{k+1},\cdots,\varphi_n^{k+1}\end{bmatrix}.$$

Repeating the process, we obtain the first identity

$$\begin{aligned} &\mathbf{res}\begin{bmatrix} \psi \mathrm{d}z\mathrm{d}t_1\cdots\mathrm{d}t_n \\ \varphi_1^{m_1} - t_1 f_1^{m_1},\cdots,\varphi_n^{m_n} - t_n f_n^{m_n}, t_1^{k+1},\cdots,t_n^{k+1}\end{bmatrix} \\ =&\ \mathbf{res}\begin{bmatrix} (f_1^{m_1}\cdots f_n^{m_n})^k\psi \mathrm{d}z \\ \varphi_1^{(k+1)m_1},\cdots,\varphi_n^{(k+1)m_n}\end{bmatrix}. \end{aligned}$$

For the second identity, we observe that

$$\begin{aligned} &\begin{bmatrix} (\varphi_1\cdots\varphi_n)^k\psi \mathrm{d}z\mathrm{d}t_1 \\ \varphi_1^{m_1}(\varphi_1^k - t_1 f_1^k), \varphi_2^{(k+1)m_2},\cdots,\varphi_n^{(k+1)m_n}, t_1^{m_1+1}\end{bmatrix} \\ =&\begin{bmatrix} (\varphi_1\cdots\varphi_n)^k\left(\sum_{i=0}^{m_1} (\varphi_1^k)^i (t_1 f_1^k)^{m_1-i}\right)\psi \mathrm{d}z\mathrm{d}t_1 \\ \varphi_1^{m_1}(\varphi_1^{(m_1+1)k} - t_1^{m_1+1} f_1^{(m_1+1)k}), \varphi_2^{(k+1)m_2},\cdots,\varphi_n^{(k+1)m_n}, t_1^{m_1+1}\end{bmatrix} \\ =&\begin{bmatrix} (\varphi_2\cdots\varphi_n)^k\left(\sum_{i=0}^{m_1} (\varphi_1^k)^i (t_1 f_1^k)^{m_1-i}\right)\psi \mathrm{d}z\mathrm{d}t_1 \\ \varphi_1^{(k+1)m_1}, \varphi_2^{(k+1)m_2},\cdots,\varphi_n^{(k+1)m_n}, t_1^{m_1+1}\end{bmatrix}. \end{aligned}$$

Applying the residue map, we get

$$\begin{aligned} &\mathbf{res}\begin{bmatrix} (\varphi_1\cdots\varphi_n)^k\psi \mathrm{d}z\mathrm{d}t_1 \\ \varphi_1^{m_1}(\varphi_1^k - t_1 f_1^k), \varphi_2^{(k+1)m_2},\cdots,\varphi_n^{(k+1)m_n}, t_1^{m_1+1}\end{bmatrix} \\ =&\ \mathbf{res}\begin{bmatrix} (f_1^{m_1}\varphi_2\cdots\varphi_n)^k\psi \mathrm{d}z \\ \varphi_1^{(k+1)m_1}, \varphi_2^{(k+1)m_2},\cdots,\varphi_n^{(k+1)m_n}\end{bmatrix}. \end{aligned}$$

Repeating the process, we obtain the second identity

$$\mathbf{res}\begin{bmatrix} (\varphi_1\cdots\varphi_n)^k\psi \mathrm{d}z\mathrm{d}t_1\cdots\mathrm{d}t_n \\ \varphi_1^{m_1}(\varphi_1^k - t_1 f_1^k),\cdots,\varphi_n^{m_n}(\varphi_n^k - t_n f_n^k), t_1^{m_1+1},\cdots,t_n^{m_n+1}\end{bmatrix}$$
$$= \mathbf{res}\begin{bmatrix} (f_1^{m_1}\cdots f_n^{m_n})^k\psi \mathrm{d}z \\ \varphi_1^{(k+1)m_1},\cdots,\varphi_n^{(k+1)m_n}\end{bmatrix}.$$

□

MacMahon's master theorem can be treated within a power series ring of n variables, say $\kappa[[z_1,\cdots,z_n]]$. The following proof taken from [21, Example 1] is an interpretation of Good's idea [13].

MacMahon's Master Theorem. Given positive integers $m_1,\cdots,m_n$ and $a_{ij}\in\kappa$, we denote $f_i=\sum_{j=1}^n a_{ij}z_j$ and $\Delta=\det(\delta_{ij}-a_{ij}x_i)$, where $x_i=z_i/(1+f_i)$ are variables. Then

$$\mathbf{res}\begin{bmatrix} f_1^{m_1}\cdots f_n^{m_n}\mathrm{d}z \\ z_1^{m_1+1},\cdots,z_n^{m_n+1}\end{bmatrix} = \mathbf{res}\begin{bmatrix} \Delta^{-1}\mathrm{d}x \\ x_1^{m_1+1},\cdots,x_n^{m_n+1}\end{bmatrix}.$$

Proof. Observe that

$$\mathbf{res}\begin{bmatrix} f_1^{m_1}\cdots f_n^{m_n}\mathrm{d}z \\ z_1^{m_1+1},\cdots,z_n^{m_n+1}\end{bmatrix} = \mathbf{res}\begin{bmatrix} (1+f_1)^{m_1}\cdots(1+f_n)^{m_n}\mathrm{d}z \\ z_1^{m_1+1},\cdots,z_n^{m_n+1}\end{bmatrix}.$$

Since

$$\mathrm{d}x_1\cdots\mathrm{d}x_n = (1+f_1)^{-1}\cdots(1+f_n)^{-1}\Delta\,\mathrm{d}z_1\cdots\mathrm{d}z_n,$$

MacMahon's master theorem stands for two ways to look at the generalized fraction

$$\begin{bmatrix} (1+f_1)^{m_1}\cdots(1+f_n)^{m_n}\mathrm{d}z \\ z_1^{m_1+1},\cdots,z_n^{m_n+1}\end{bmatrix} = \begin{bmatrix} \Delta^{-1}\mathrm{d}x_1\cdots\mathrm{d}x_n \\ x_1^{m_1+1},\cdots,x_n^{m_n+1}\end{bmatrix}$$

by residues from variables $z_1,\cdots,z_n$ and from variables $x_1,\cdots,x_n$. □

6.5.4 Dyson's conjecture

Let $a_1,\cdots,a_n$ be non-negative integers. Dyson's conjecture [6], that the constant term of

$$\prod_{1\le i\ne j\le n}\left(1-\frac{X_i}{X_j}\right)^{a_i}$$

is equal to

$$\frac{(a_1+\cdots+a_n)!}{a_1!\cdots a_n!},$$

was proved by Wilson [35] and Gunson [17]. We give two proofs. The first proof taken from [19] is due to Wilson and involves changes of parameters. The second proof due to Good [14] doesn't involve changes of parameters.

Let $X_1, \cdots, X_n$ be variables of $\mathbb{Q}[[e^{\mathbb{Z}^n}]]$ over $\mathbb{Q} = \mathbb{Q}[[e^0]]$. We assume that $\log X_1 > \cdots > \log X_n$. Let

$$\Phi_i = \prod_{j=1, j\neq i}^{n} \frac{X_j}{X_j - X_i} \in \mathbb{Q}[[e^{\mathbb{Z}^n}]].$$

Using Lagrange interpolation, one shows that $\Phi_1, \cdots, \Phi_n$ satisfy the relation

$$\Phi_1 + \cdots + \Phi_n = 1.$$

Wilson's proof is based on the parameters $X_1, \Phi_2, \cdots, \Phi_n$, whose multiplicities with respect to $X_1, \cdots, X_n$ have determinant

$$\det \begin{pmatrix} 1 & 0 & 0 & \cdots & 0 \\ 1 & -1 & 0 & \cdots & 0 \\ 1 & 1 & -2 & \cdots & 0 \\ \vdots & \vdots & \vdots & \ddots & \vdots \\ 1 & 1 & 1 & \cdots & -(n-1) \end{pmatrix} = (n-1)!(-1)^{n-1}.$$

For representations in terms of $\mathrm{dlog}\,\mathbf{X}$, we note that

$$\frac{\partial \log \Phi_i}{\partial \log X_j} = \begin{cases} -\sum_{k=1, k\neq i}^{n} \frac{X_i}{X_i - X_k}, & \text{if } i = j; \\ \frac{X_i}{X_i - X_j}, & \text{if } i \neq j. \end{cases}$$

It is shown in the proof of [35, Lemma 3] that the $(n-1) \times (n-1)$ matrix $((\partial \log \Phi_i)/(\partial \log X_j))_{2 \leq i,j \leq n}$ has determinant $c(n-1)!(-1)^{n-1}\Phi_1$ for some $c \in \mathbb{Q}$. Therefore

$$\mathrm{dlog}\,X_1 \wedge \mathrm{dlog}\,\Phi_2 \wedge \cdots \wedge \mathrm{dlog}\,\Phi_n = c(n-1)!(-1)^{n-1}\Phi_1 \,\mathrm{dlog}\,\mathbf{X}.$$

The scalar c is not zero, since $\mathrm{dlog}\,X_1 \wedge \mathrm{dlog}\,\Phi_2 \wedge \cdots \wedge \mathrm{dlog}\,\Phi_n$ generates $\wedge^n \Omega_{\mathscr{G}/\mathscr{H}}$. Let

$$\Psi(X_1, \cdots, X_n) = X_2^{-a_2} \cdots X_n^{-a_n} \sum_{k=0}^{\infty} \binom{k + a_1 - 1}{k} (X_2 + \cdots + X_n)^k.$$

Since $\Phi_2, \cdots, \Phi_n$ are positive, we have a representation

$$\Phi_1^{-a_1} \cdots \Phi_n^{-a_n} = \Psi(X_1, \Phi_2, \cdots, \Phi_n).$$

What we need to compute is the constant term of $\Psi(X_1, \Phi_2, \cdots, \Phi_n)$, that is, the residue of

$$\begin{bmatrix} \Psi(X_1, \Phi_2, \cdots, \Phi_n)\,\mathrm{dlog}\,\mathbf{X} \\ \log \mathbf{X} \end{bmatrix}$$

$$= \begin{bmatrix} \frac{c^{-1}\Psi(X_1, \Phi_2, \cdots, \Phi_n)}{1 - \Phi_2 - \cdots - \Phi_n} \,\mathrm{dlog}\,X_1 \wedge \mathrm{dlog}\,\Phi_2 \wedge \cdots \wedge \mathrm{dlog}\,\Phi_n \\ \log X_1, \log \Phi_2, \cdots, \log \Phi_n \end{bmatrix}.$$

By Jacobi law, the constant term of $\Psi(X_1,\Phi_2,\cdots,\Phi_n)$ is the same as that of

$$\frac{c^{-1}\Psi(X_1,\cdots,X_n)}{1-X_2-\cdots-X_n}=\frac{c^{-1}}{X_2^{a_2}\cdots X_n^{a_n}}\sum_{k=0}^{\infty}\binom{k+a_1}{a_1}(X_2+\cdots+X_n)^k,$$

which is

$$c^{-1}\binom{a_1+\cdots+a_n}{a_1}\binom{a_2+\cdots+a_n}{a_2,\cdots,a_n}=c^{-1}\frac{(a_1+\cdots+a_n)!}{a_1!\cdots a_n!}$$

occurring when $k=a_2+\cdots+a_n$. Now Dyson's conjecture for the trivial case $a_0=\cdots=a_n=0$ shows $c=1$.

For the second proof, we consider

$$D(a_1,\cdots,a_n):=\prod_{1\le i\ne j\le n}(X_j-X_i)^{a_i}.$$

What we need to show is the identity

$$\mathbf{res}\begin{bmatrix}\dfrac{X_1^{a_1}\cdots X_n^{a_n}D(a_1,\cdots,a_n)}{(X_1\cdots X_n)^{a_1+\cdots+a_n}}\,\mathrm{dlog}\,\mathbf{X}\\ \log\mathbf{X}\end{bmatrix}=\frac{(a_1+\cdots+a_n)!}{a_1!\cdots a_n!}.$$

If all a_i are positive, multiply the identity $\sum_{i=1}^{n}\Phi_i=1$ by $D(a_1,\cdots,a_n)$, we get

$$\sum_{i=1}^{n}X_1\cdots\hat{X}_i\cdots X_nD(a_1,\cdots,a_{i-1},a_i-1,a_{i+1},\cdots,a_n)=D(a_1,\cdots,a_n).$$

Now we prove Dyson's conjecture by induction on the number $|a|:=a_1+\cdots+a_n$. If $|a|=1$, Dyson's conjecture can be checked directly. Assume $|a|>1$ and Dyson's conjecture holds for smaller $|a|$. We may assume that all a_i are positive, then

$$\begin{aligned}&\mathbf{res}\begin{bmatrix}\dfrac{X_1^{a_1}\cdots X_n^{a_n}D(a_1,\cdots,a_n)}{(X_1\cdots X_n)^{a_1+\cdots+a_n}}\,\mathrm{dlog}\,\mathbf{X}\\ \log\mathbf{X}\end{bmatrix}\\ &=\sum_{i=1}^{n}\mathbf{res}\begin{bmatrix}\dfrac{X_1^{a_1}\cdots X_i^{a_i-1}\cdots X_n^{a_n}D(a_1,\cdots,a_i-1,\cdots,a_n)}{(X_1\cdots X_n)^{a_1+\cdots+a_n-1}}\,\mathrm{dlog}\,\mathbf{X}\\ \log\mathbf{X}\end{bmatrix}\\ &=\sum_{i=1}^{n}\frac{(a_1+\cdots+a_n-1)!}{a_1!\cdots a_{i-1}!(a_i-1)!a_{i+1}!\cdots a_n!}\\ &=\frac{(a_1+\cdots+a_n)!}{a_1!\cdots a_n!}.\end{aligned}$$

The second proof can be also written in terms of local cohomology residues. See [23, Identity 14].

6.5.5 Constraints of Analytic Functions

Using analytic differentials, one needs to take care of the issue of convergence. We provide an example from [9, p.127 - p.129] to show how analytic procedures can be transformed to algebraic language so that convergence becomes irrelevant. We work on the power series ring $\mathbb{Q}[[u,v]]$.

Let S_n be the number of non-isomorphic tournaments with n vertices having a unique Hamiltonian circuit. It is known that $S_4 = 1$ and for $n \geq 5$

$$S_n = 1 + \sum_{k=1}^{n-3} \sum_{p=0}^{\min(k-1,n-k-3)} 2^{n-k-p-4} \,\mathbf{res}\left[\frac{(1+v)^{k-1}(2+uv)\mathrm{d}u\mathrm{d}v}{(1-u)^{p+1}u^{n-k-2-p}, v^{p+2}}\right].$$

We will show that

$$S_n = \mathbf{res}\left[\frac{\mathrm{d}w}{(w^2-3w+1)w^{n-3}}\right] \tag{6.5}$$

for $n \geq 4$.

First, with some terms vanishing, we rewrite

$$S_n = 1 + \sum_{k=1}^{n-3} 2^{n-k-3} \sum_{p=0}^{\infty} \mathbf{res}\left[\frac{u^p(1+v)^{k-1}(2+uv)\mathrm{d}u\mathrm{d}v}{2^p(1-u)^p u^{n-k-1}, v^{p+1}}\right]$$

by changing the index p. Let $y = 2^{-1}(1-u)^{-1}u$ and $z = v - y$. Then $\mathrm{d}u\mathrm{d}v = \mathrm{d}u\mathrm{d}z$ and

$$\begin{aligned} S_n &= 1 + \sum_{k=1}^{n-3} 2^{n-k-3} \,\mathbf{res}\left[\frac{(1+z+y)^{k-1}(2+uz+uy)\mathrm{d}u\mathrm{d}z}{u^{n-k-1}, z}\right] \\ &= 1 + \sum_{k=1}^{n-3} 2^{n-k-3} \,\mathbf{res}\left[\frac{(1+y)^{k-1}(2+uy)\mathrm{d}u}{u^{n-k-1}}\right] \\ &= 1 + \sum_{k=1}^{n-3} 2^{n-2k-3} \,\mathbf{res}\left[\frac{(1-u)^{-k}(2-u)^{k+1}\mathrm{d}u}{u^{n-k-1}}\right]. \end{aligned}$$

Note that

$$\mathbf{res}\left[\frac{(1-u)^{-k}(2-u)^{k+1}\mathrm{d}u}{u^{n-k-1}}\right] = \begin{cases} 2^{-n+2k+3}, & \text{if } k = n-2; \\ 0, & \text{if } k > n-2. \end{cases}$$

Hence

$$\begin{aligned} S_n &= \sum_{k=1}^{\infty} 2^{n-2k-3} \,\mathbf{res}\left[\frac{(1-u)^{-k}(2-u)^{k+1}\mathrm{d}u}{u^{n-k-1}}\right] \\ &= 2^{n-3} \,\mathbf{res}\left[\frac{(2-u)\sum_{k=1}^{\infty} 4^{-k}(1-u)^{-k}u^k(2-u)^k\mathrm{d}u}{u^{n-1}}\right] = \end{aligned}$$

$$= 2^{n-3}\,\mathbf{res}\left[\frac{4^{-1}(1-u)^{-1}u(2-u)^2\mathrm{d}u}{(1-4^{-1}(1-u)^{-1}u(2-u))u^{n-1}}\right]$$
$$= 2^{n-3}\,\mathbf{res}\left[\frac{(2-u)^2\mathrm{d}u}{(u^2-6u+4)u^{n-2}}\right].$$

We remark that the analytic constraint

$$|4^{-1}(1-u)^{-1}u(2-u)| < 1$$

in a closed domain $|u| \leq \rho$ for the summation $\sum 4^{-k}(1-u)^{-k}u^k(2-u)^k$ is irrelevant in our computation. Finally, the change of variable $w = u/2$ gives rise to (6.5).

Appendix: Calculus with/without Residues

Local cohomology residues (and their sums running over local rings of a variety) may serve as an operational tool as provided by analytic residues or other algebraic residues. In the appendix, we revisit the key tool (from the computational point of view) for effective Nullstellensatz [2], which uses residues via Hochschild homology [28] to replace analytic residues in previous work. We also provide an algebraic analogue of an analytic formula of Weil [34] in terms of local cohomology residues. We remark that another algebraic analogue of Weil's formula have been already given in residues via Hochschild homology [4].

Our viewpoint is that some versions of transformation formula in other theories of residues are in fact properties of generalized fractions in local cohomology modules. We begin with a technical lemma, whose proof uses Koszul complexes (see [5, §1.6]).

Lemma 6.1. *Let $f_1, \cdots, f_n$ be a regular sequence in a commutative ring R and $g_1, \cdots, g_n$ be another sequence in R with relations*

$$g_i = \sum_{j=1}^{n} a_{ij} f_j = \sum_{j=1}^{n} b_{ij} f_j, \tag{6.6}$$

where $a_{ij}, b_{ij} \in R$. Then $\det(a_{ij}) - \det(b_{ij}) \in (g_1, \cdots, g_n)$.

Proof. The relations (6.6) determine two morphisms of complexes

$$\phi_1, \phi_2\colon K_\bullet(g_1, \cdots, g_n) \to K_\bullet(f_1, \cdots, f_n)$$

of Koszul complexes, whose restrictions $K_n(g_1, \cdots, g_n) \to K_n(f_1, \cdots, f_n)$ are multiplications by $\det(a_{ij})$ and by $\det(b_{ij})$, respectively. As the sequence $f_1, \cdots, f_n$ is regular, the complex $K_\bullet(f_1, \cdots, f_n)$ provides a free resolution of $R/(f_1, \cdots, f_n)$, and hence morphisms ϕ_1 and ϕ_2 are homotopic. The morphisms

$$K^\bullet(f_1, \cdots, f_n) \to K^\bullet(g_1, \cdots, g_n)$$

induced by ϕ_1 and ϕ_2 by taking duals are also homotopic. Therefore multiplications by $\det(a_{ij})$ and by $\det(b_{ij})$ give rise to the same map

$$H^n(f_1,\cdots,f_n) \to H^n(g_1,\cdots,g_n).$$

By self-duality of Koszul complexes,

$$H^n(f_1,\cdots,f_n) = H_0(f_1,\cdots,f_n) = \frac{R}{(f_1,\cdots,f_n)},$$
$$H^n(g_1,\cdots,g_n) = H_0(g_1,\cdots,g_n) = \frac{R}{(g_1,\cdots,g_n)}.$$

We get $\det(a_{ij}) - \det(b_{ij}) \in (g_1,\cdots,g_n)$. □

The following transformation law holds without residues, *cf.* [2, Proposition 2.2].

Proposition 6.1. *Let $f_0, f_1,\cdots,f_n$ and $f_0, g_1,\cdots,g_n$ be two regular sequences of $\kappa[[X_0,X_1,\cdots,X_n]]$. Assume that there are $s_1,\cdots,s_n \in \mathbb{N}$ and $a_{j\ell} \in \kappa[[X_0,X_1,\cdots,X_n]]$ such that*

$$f_0^{s_j} g_j = \sum_{\ell=1}^{n} a_{j\ell} f_\ell, \qquad j = 1,\cdots,n.$$

Then, for any $k_0 \in \mathbb{N}$ and $Q \in \kappa[[X_0,X_1,\cdots,X_n]]$, one has

$$\begin{bmatrix} Q\mathrm{d}X_0\cdots\mathrm{d}X_n \\ f_0^{k_0}, f_1,\cdots,f_n \end{bmatrix} = \begin{bmatrix} Q\det(a_{j\ell})\mathrm{d}X_0\cdots\mathrm{d}X_n \\ f_0^{k_0+s_1+\cdots+s_n}, g_1,\cdots,g_n \end{bmatrix}.$$

Proof. $f_1,\cdots,f_n,f_0^{s_j}$ is a regular sequence for any $1 \le j \le n$. So there exist $b_{j1},\cdots,b_{jn} \in \kappa[[X_0,X_1,\cdots,X_n]]$ such that

$$g_j = \sum_{\ell=1}^{n} b_{j\ell} f_\ell.$$

By Lemma,

$$f_0^{s_1+\cdots+s_n}\det(b_{j\ell}) - \det(a_{j\ell}) = \det(f_0^{s_j} b_{j\ell}) - \det(a_{j\ell}) \in (f_0^{s_1} g_1,\cdots,f_0^{s_n} g_n).$$

Therefore

$$\begin{bmatrix} Q\mathrm{d}X_0\cdots\mathrm{d}X_n \\ f_0^{k_0}, f_1,\cdots,f_n \end{bmatrix} = \begin{bmatrix} Q\det(b_{j\ell})\mathrm{d}X_0\cdots\mathrm{d}X_n \\ f_0^{k_0}, g_1,\cdots,g_n \end{bmatrix} = \begin{bmatrix} Q\det(a_{j\ell})\mathrm{d}X_0\cdots\mathrm{d}X_n \\ f_0^{k_0+s_1+\cdots+s_n}, g_1,\cdots,g_n \end{bmatrix}.$$

□

Next proposition as a variant of transformation law is again a property of generalized fractions, *cf.* [2, Proposition 2.3]. See also [3, 27]. Our proof follows [2, Proposition 2.3] with its multi-index notation.

Proposition 6.2. *Let $f=(f_1,\dots,f_n)$ and $g=(g_1,\dots,g_n)$ be two regular sequences in $\kappa[[X_1,\cdots,X_n]]$ with relations*

$$g_j=\sum_{\ell=1}^{n} a_{j\ell} f_\ell \qquad (j=1,\dots,n).$$

where $a_{j\ell}\in\kappa[[X_1,\cdots,X_n]]$ and we let Δ be the determinant of the matrix $A:=(a_{j\ell})$. Then for any $Q\in\kappa[[X_1,\cdots,X_n]]$ and any $k\in\mathbb{N}^n$, we have

$$\begin{bmatrix} Q\mathrm{d}X \\ f^{k+1}\end{bmatrix}=\sum_{|q_{;i}|=k_i\forall i}\left(\prod_{\ell=1}^{n}\binom{\mu_\ell}{q_{\ell;}}\right)\begin{bmatrix} Q\Delta\prod_{i,j=1}^{n}(a_{ji})^{q_{ji}}\mathrm{d}X \\ g^{\mu+1}\end{bmatrix}.$$

where

$$q_{;j}=(q_{1j},\dots,q_{nj}),\ q_{i;}=(q_{i1},\dots,q_{in}),\ \mu_i=|q_{i;}|$$

and

$$\binom{\mu_i}{q_{i;}}=\frac{\mu_i}{q_{i1}!\dots q_{in}!}.$$

Proof. We work on the power series ring $\kappa[[U_1,\cdots,U_n,X_1,\cdots,X_n]]$.

$$\begin{aligned}
\begin{bmatrix} Q\mathrm{d}X \\ f^{k+1}\end{bmatrix} &= \mathbf{res}_U\begin{bmatrix} Q\prod_{j=1}^{n}(f_j^{k_j}+f_j^{k_j-1}U_j+\cdots+U_j^{k_j})\mathrm{d}U\mathrm{d}X \\ U^{k+1},f^{k+1}\end{bmatrix}\\
&= \mathbf{res}_U\begin{bmatrix} Q\prod_{j=1}^{n}(f_j^{k_j+1}-U_j^{k_j+1})\mathrm{d}U\mathrm{d}X \\ U^{k+1},(f-U)f^{k+1}\end{bmatrix}\\
&= \mathbf{res}_U\begin{bmatrix} Q\prod_{j=1}^{n}f_j^{k_j+1}dU\mathrm{d}X \\ U^{k+1},(f-U)f^{k+1}\end{bmatrix}\\
&= \mathbf{res}_U\begin{bmatrix} Q\mathrm{d}U\mathrm{d}X \\ U^{k+1},f-U\end{bmatrix}
\end{aligned}$$

Using the relation $A(f-U)=g-AU$ and the above identity, we can write

$$\begin{bmatrix} Q\mathrm{d}X \\ f^{k+1}\end{bmatrix}=\mathbf{res}_U\begin{bmatrix} Q\Delta\mathrm{d}U\mathrm{d}X \\ U^{k+1},g-AU\end{bmatrix}.$$

As $(AU)_j^{|k|+1}$ is contained in the ideal generated by $U_1^{k_1+1},\cdots,U_n^{k_n+1}$ for each j,

$$\begin{aligned}
\begin{bmatrix} Q\mathrm{d}X \\ f^{k+1}\end{bmatrix} &= \mathbf{res}_U\begin{bmatrix} Q\Delta\prod_{j=1}^{n}(\sum_{\tau_j=0}^{|k|}g_j^{|k|-\tau_j}(AU)_j^{\tau_j})\mathrm{d}U\mathrm{d}X \\ U^{k+1},g^{|k|+1}-(AU)^{|k|+1}\end{bmatrix}\\
&= \mathbf{res}_U\begin{bmatrix} Q\Delta\prod_{j=1}^{n}(\sum_{\tau_j=0}^{|k|}g_j^{|k|-\tau_j}(AU)_j^{\tau_j})\mathrm{d}U\mathrm{d}X \\ U^{k+1},g^{|k|+1}\end{bmatrix}.
\end{aligned}$$

The product $\prod_{j=1}^{n}(\sum_{\tau_j=0}^{|k|} g_j^{|k|-\tau_j}(AU)_j^{\tau_j})$ is a sum of the monomials of the form

$$\left(\prod_{j=1}^{n} g_j^{|k|-\tau_j}\right)\left(\prod_{i,j=1}^{n}(a_{ji}U_i)^{q_{ji}}\right)$$

satisfying the condition $\tau_j = |q_{j;}| = \mu_j$. There are $\prod_{\ell=1}^{n}\binom{\mu_\ell}{q_{\ell;}}$ such monomials in the product. Therefore

$$\begin{aligned}
&\begin{bmatrix} Q\mathrm{d}X \\ f^{k+1} \end{bmatrix} \\
&= \sum_{\mu_j \le |k|} \mathbf{res}_U \begin{bmatrix} Q\Delta\left(\prod_{\ell=1}^{n}\binom{\mu_\ell}{q_{\ell;}}\right)\left(\prod_{j=1}^{n} g_j^{|k|-\mu_j}\right)\left(\prod_{i,j=1}^{n}(a_{ji}U_i)^{q_{ji}}\right)\mathrm{d}U\mathrm{d}X \\ U^{k+1}, g^{|k|+1} \end{bmatrix} \\
&= \sum_{\mu_j \le |k|} \left(\prod_{\ell=1}^{n}\binom{\mu_\ell}{q_{\ell;}}\right) \mathbf{res}_U \begin{bmatrix} Q\Delta\prod_{i,j=1}^{n}(a_{ji}U_i)^{q_{ji}}\mathrm{d}U\mathrm{d}X \\ U^{k+1}, g^{\mu+1} \end{bmatrix}.
\end{aligned}$$

In order to get non-trivial residues, q_{ji} has to satisfy the condition $|q_{;i}| = k_i$. With this condition, $\mu_j \le |k|$ is automatically satisfied. Therefore

$$\begin{bmatrix} Q\mathrm{d}X \\ f^{k+1} \end{bmatrix} = \sum_{|q_{;i}|=k_i \forall i} \left(\prod_{\ell=1}^{n}\binom{\mu_\ell}{q_{\ell;}}\right) \begin{bmatrix} Q\Delta\prod_{i,j=1}^{n}(a_{ji})^{q_{ji}}\mathrm{d}X \\ g^{\mu+1} \end{bmatrix}.$$

□

Weil's formula [1, Theorem 9.1 and Theorem 24.9], [18], [34] can be stated in terms of local cohomology residues as follows. See also [4] for another algebraic treatment.

Proposition 6.3. *Let $f_1, \cdots, f_n$ be a regular sequence of $\kappa[[X_1, \cdots, X_n]]$. Assume that*

$$f_i(\zeta_1, \cdots, \zeta_n) - f_i(X_1, \cdots, X_n) = \sum_{j=1}^{n}(\zeta_j - X_j)g_{ij} \quad (1 \le j \le n).$$

for $g_{ij} \in \kappa[[\zeta_1, \cdots, \zeta_n, X_1, \cdots, X_n]]$. Let Δ be the determinant of the matrix (g_{ij}). Then, for any $h \in \kappa[[X_1, \cdots, X_n]]$ and any $j \in \mathbb{N}^n$,

$$\begin{aligned}
\mathbf{res}\begin{bmatrix} h\mathrm{d}X \\ X^j \end{bmatrix} &= \mathbf{res}\begin{bmatrix} h(\zeta)\Delta\mathrm{d}\zeta\mathrm{d}X \\ f(\zeta) - f(X), X^j \end{bmatrix} \\
&= \sum_{i \in \mathbb{N}^n} \mathbf{res}\begin{bmatrix} h(\zeta)\Delta f_1^{i_1-1}(X)\cdots f_n^{i_n-1}(X)\mathrm{d}\zeta\mathrm{d}X \\ f^i(\zeta), X^j \end{bmatrix}.
\end{aligned}$$

Proof. To show the first identity, we may assume that h is a monomial $X_1^{\ell_1}\cdots X_n^{\ell_n}$. By the transformation law,

$$\mathbf{res}\begin{bmatrix} \zeta_1^{\ell_1}\cdots\zeta_n^{\ell_n}\Delta \mathrm{d}\zeta \mathrm{d}X \\ f(\zeta)-f(X),X^j \end{bmatrix} = \mathbf{res}\begin{bmatrix} \zeta_1^{\ell_1}\cdots\zeta_n^{\ell_n}\mathrm{d}\zeta \mathrm{d}X \\ \zeta-X,X^j \end{bmatrix}.$$

We may replace ζ_i by X_i. For instance,

$$\begin{aligned}\mathbf{res}\begin{bmatrix} \zeta_1^{\ell_1}\cdots\zeta_n^{\ell_n}\mathrm{d}\zeta \mathrm{d}X \\ \zeta-X,X^j \end{bmatrix} &= \mathbf{res}\begin{bmatrix} (X_1+\zeta_1-X_1)\zeta_1^{\ell_1-1}\zeta_2^{\ell_2}\cdots\zeta_n^{\ell_n}\mathrm{d}\zeta \mathrm{d}X \\ \zeta-X,X^j \end{bmatrix} \\ &= \mathbf{res}\begin{bmatrix} X_1\zeta_1^{\ell_1-1}\zeta_2^{\ell_2}\cdots\zeta_n^{\ell_n}\mathrm{d}\zeta \mathrm{d}X \\ \zeta-X,X^j \end{bmatrix}.\end{aligned}$$

Therefore

$$\mathbf{res}\begin{bmatrix} \zeta_1^{\ell_1}\cdots\zeta_n^{\ell_n}\mathrm{d}\zeta \mathrm{d}X \\ \zeta-X,X^j \end{bmatrix} = \mathbf{res}\begin{bmatrix} X_1^{\ell_1}\cdots X_n^{\ell_n}\mathrm{d}(\zeta-X)\mathrm{d}X \\ \zeta-X,X^j \end{bmatrix} = \mathbf{res}\begin{bmatrix} X_1^{\ell_1}\cdots X_n^{\ell_n}\mathrm{d}X \\ X^j \end{bmatrix}.$$

For the second identity, we choose ℓ such that $f_1^\ell,\cdots,f_n^\ell \in (X_1^{j_1},\cdots,X_n^{j_n})$. Then

$$\begin{aligned}&\mathbf{res}\begin{bmatrix} h(\zeta)\Delta \mathrm{d}\zeta \mathrm{d}X \\ f(\zeta)-f(X),X^j \end{bmatrix} \\ &= \mathbf{res}\begin{bmatrix} h(\zeta)\Delta(f_1^\ell(\zeta)-f_1^\ell(X))\cdots(f_n^\ell(\zeta)-f_n^\ell(X))\mathrm{d}\zeta \mathrm{d}X \\ f^\ell(\zeta)(f(\zeta)-f(X)),X^j \end{bmatrix} \\ &= \sum_{i_1,\cdots,i_n=1}^{\ell} \mathbf{res}\begin{bmatrix} h(\zeta)\Delta(f_1^{\ell-i_1}(\zeta)f_1^{i_1-1}(X))\cdots(f_n^{\ell-i_n}(\zeta)f_n^{i_n-1}(X))\mathrm{d}\zeta \mathrm{d}X \\ f^\ell(\zeta),X^j \end{bmatrix} \\ &= \sum_{i_1,\cdots,i_n>0} \mathbf{res}\begin{bmatrix} h(\zeta)\Delta f_1^{i_1-1}(X)\cdots f_n^{i_n-1}(X)\mathrm{d}\zeta \mathrm{d}X \\ f^i(\zeta),X^j \end{bmatrix}.\end{aligned}$$

□

References

1. I. A. Aĭzenberg and A. P. Yuzhakov. *Integral representations and residues in multidimensional complex analysis*. American Mathematical Society, Providence, R.I., 1983.
2. C. A. Berenstein and A. Yger. Residue calculus and effective Nullstellensatz. *Amer. J. Math.*, 121(4):723–796, 1999.
3. J.-Y. Boyer and M. Hickel. Une généralisation de la loi de transformation pour les résidus. *Bull. Soc. Math. France*, 125(3):315–335, 1997.
4. J.-Y. Boyer and M. Hickel. Extension dans un cadre algébrique d'une formule de Weil. *Manuscripta Math.*, 98(2):195–223, 1999.
5. W. Bruns and J. Herzog. *Cohen-Macaulay Rings*. Cambridge University Press, 1993.
6. F. J. Dyson. Statistical theory of the energy levels of complex systems. I. *J. Mathematical Phys.*, 3:140–156, 1962.
7. G. P. Egorychev. Inversion of one-dimensional combinatorial relations. In *Some questions on the theory of groups and rings (Russian)*, pages 110–122, 176. Inst. Fiz. im. Kirenskogo Sibirsk. Otdel. Akad. Nauk SSSR, Krasnoyarsk, 1973.

8. G. P. Egorychev. The inversion of combinatorial relations. *Kombinatornyĭ Anal.*, (Vyp. 3):10–14, 1974.
9. G. P. Egorychev. *Integral Representation and the Computation of Combinatorial Sums*, volume 59 of *Translation of Mathematical Monographs*. American Mathematical Society, 1984.
10. G. P. Egorychev and E. V. Zima. Decomposition and group theoretic characterization of pairs of inverse relations of the Riordan type. *Acta Appl. Math.*, 85(1-3):93–109, 2005.
11. G. P. Egorychev and E. V. Zima. Integral representation and algorithms for closed form summation. Handbook of Algebra, vol. 5, (ed. M. Hazewinkel), Elsevier, 459–529, 2008.
12. I. M. Gessel. A combinatorial proof of the multivaribles Lagrange inversion formula. *Journal of Combinatorial Theory, Series A*, 45:178–195, 1987.
13. I. J. Good. A short proof of MacMahon's 'master theorem'. *Proc. Cambridge Philos. Soc.*, 58:160, 1962.
14. I. J. Good. Short proof of a conjecture by Dyson. *J. Mathematical Phys.*, 11:1884, 1970.
15. I. P. Goulden and D. M. Jackson. *Combinatorial Enumeration*. John Wiley & Sons, 1983.
16. P. Griffiths and J. Harris. *Principles of Algebraic Geometry*. John Wiley & Sons, Inc., 1978.
17. J. Gunson. Proof of a conjecture by Dyson in the statistical theory of energy levels. *J. Mathematical Phys.*, 3:752–753, 1962.
18. M. Hervé. Intégrale d'André Weil. In *Séminaire H. Cartan de l'Ecole Normale Supérieure: Fonction analytiques de plusieurs variables complexes t. 4*. 1951–52. Exposé 6.
19. I-C. Huang. Changes of parameters for generalized power series. *Comm. Algebra* (in print).
20. I-C. Huang. Pseudofunctors on modules with zero dimensional support. *Mem. Amer. Math. Soc.*, 114(548):xii+53, 1995.
21. I-C. Huang. Applications of residues to combinatorial identities. *Proc. Amer. Math. Soc.*, 125(4):1011–1017, 1997.
22. I-C. Huang. Reversion of power series by residues. *Comm. Algebra*, 26(3):803–812, 1998.
23. I-C. Huang. Residue methods in combinatorial analysis. In *Local Cohomology and its Applications*, volume 226 of *Lecture Notes in Pure and Appl. Math.*, pages 255–342. Marcel Dekker, 2001.
24. I-C. Huang. Inverse relations and Schauder bases. *J. Combin. Theory Ser. A*, 97(2):203–224, 2002.
25. C. G. I. Jacobi. De resolutione aequationum per series infinitas. *J. Reine Angew. Math.*, 6:257–286, 1830.
26. E. Kunz. *Kähler Differentials*. Vieweg, Braunschweig, Wiesbaden, 1986.
27. A. M. Kytmanov. A formula for the transformation of the Grothendieck residue and some of its applications. *Sibirsk. Mat. Zh.*, 29(3):198–202, 223, 1988.
28. J. Lipman. *Residues and traces of differential forms via Hochschild homology*, volume 61. Contemporary Mathematics of the AMS, 1987.
29. H. Matsumura. *Commutative Ring Theory*. Cambridge University Press, 1986.
30. D. Merlini, R. Sprugnoli, and M. C. Verri. Lagrange inversion: when and how. *Acta Appl. Math.*, 94(3):233–249 (2007), 2006.
31. D. Merlini, R. Sprugnoli, and M. C. Verri. Combinatorial sums and implicit Riordan arrays. to appear in Discrete Mathematics, doi:10.1016/j.disc.2007.12.039.
32. D. S. Passman. *The algebraic structure of group rings*. Wiley-Interscience [John Wiley & Sons], New York, 1977. Pure and Applied Mathematics.
33. J. Riordan. *Combinatorial Identities*. Wiley, 1968.
34. A. Weil. L'intégrale de Cauchy et les fonctions de plusieurs variables. *Math. Ann.*, 111(1):178–182, 1935.
35. K. G. Wilson. Proof of a conjecture by Dyson. *J. Mathematical Phys.*, 3:1040–1043, 1962.

Chapter 7
Henrici's Friendly Monster Identity Revisited

Peter Paule

Dedicated to Professor Georgy Egorychev on the occasion of his 70th birthday

Abstract We revisit Peter Henrici's friendly monster identity to present a case study on Egorychev's method. Connections to various computer algebra approaches are drawn.

7.1 Introduction

Let us consider the "bonus problem" 5.94 in [4]: Show that if $w = e^{2\pi i/3}$ we have

$$\sum_{k+l+m=3n} \left(\frac{(3n)!}{k!l!m!} \right)^2 w^{m-l} = \frac{(4n)!}{n!n!(2n)!} \qquad (n \geq 0). \tag{7.1}$$

In the "Answers to Exercises" [4, p. 526] one finds that this is a consequence of Henrici's "friendly monster" identity [7, p. 118]. Namely, set $c = 1$ and compare the coefficients of x^{3n} on both sides of

$$\begin{aligned} f(c,x)f(c,wx)f(c,w^2x) = \sum_{j=0}^{\infty} & \frac{(\frac{1}{2}c - \frac{1}{4})_j(\frac{1}{2}c + \frac{1}{4})_j}{(\frac{1}{3}c)_j(\frac{1}{3}c + \frac{1}{3})_j(\frac{1}{3}c + \frac{2}{3})_j} \\ & \times \frac{(\frac{4x}{9})^{3j}}{(\frac{2}{3}c - \frac{1}{3})_j(\frac{2}{3}c)_j(\frac{2}{3}c + \frac{1}{3})_j(c)_j j!}, \end{aligned} \tag{7.2}$$

where

Peter Paule
Research Institute for Symbolic Computation (RISC)
Johannes Kepler University, A-4040 Linz, Austria, e-mail: ppaule@risc.uni-linz.ac.at
Partially supported by SFB grant F1305 of the Austrian Science Foundation FWF.

I.S. Kotsireas, E.V. Zima (eds.), *Advances in Combinatorial Mathematics*,
DOI 10.1007/978-3-642-03562-3_7,

$$f(c,x) \doteqdot \sum_{j\geq 0} \frac{x^j}{(c)_j j!} \tag{7.3}$$

with $(c)_j = c(c+1)\cdots(c+j-1)$ if $j \geq 1$, and $(c)_0 = 1$. In addition, it is stated that "If we replace $3n$ by $3n+1$ or $3n+2$, the given sum is zero."

Remark. The right hand side of (7.2) is a hypergeometric ${}_2F_7$ series; $f(c,x)$ is a ${}_0F_1$ series which is a variant of a Bessel function of first kind.

In the second edition [5, p. 546] one finds an alternative solution that has been provided by the author of this note. In this alternative solution the way of simplifying the original multiple sum has been strongly inspired by *Egorychev's method*, treated extensively in the monograph [3]. In this note, we present a more detailed account of this solution; in addition, we relate Egorychev's method to some recent developments in computer algebra.

In Section 7.2 we use Egorychev's method to reduce the problem to a single sum (Sect. 7.2.1) which then is simplified by computer algebra (Sect. 7.2.2.1) and, alternatively, by classical hypergeometric methods (Sect. 7.2.2.2). Using the software package `MultiSum`, in Section 7.3 we simplify the friendly monster identity by a direct computer algebra attack on the originally given form. In Section 7.4 we use the package `GeneratingFunctions` to present an alternative computer algebra solution. Finally, in Section 7.5 we conclude with a discussion of the methods presented.

7.2 Egorychev's Method in Action

To cover the cases $3n$, $3n+1$, and $3n+2$ in one stroke, we define for $N \geq 0$,

$$S(N) \doteqdot \sum_{k,l} \left(\frac{N!}{k!l!(N-k-l)!} \right)^2 w^{(N-k-l)-l}.$$

The multiple sum in (7.1) is the case $N = 3n$. To simplify $S(N)$ we do not need to invoke the full power of Egorychev's method exploiting residue calculus on integral representations. As it turns out the usage of the residue functional res_z on series expansions will be sufficient.

Definition. Let $f(z) = \sum_{k=-m}^{\infty} f_n z^n$ be a (formal) Laurent series for some integer m. Then the residue functional $\mathrm{res}_z f(z)$ takes the coefficient of z^{-1} in the series expansion of $f(z)$, i.e.,

$$\operatorname*{res}_z f(z) \doteqdot f_{-1}.$$

An addition, for later convenience we also define the constant term functional,

$$\langle z^0 \rangle f(z) \doteqdot f_0.$$

7.2.1 Reduction to a single sum

We begin with a slight rewriting:

$$S(N) = \sum_{k,l} \left(\frac{N!}{k!l!(N-k-l)!} \right)^2 w^{N-k+l} = \sum_{k,l} \binom{N}{k}^2 \binom{k}{l}^2 w^{l+k}. \tag{7.4}$$

The last equality is obtained by replacing k with $N-k$. After this preprocessing step we reduce with Egorychev's method. Rewriting the inner sum at the right hand side as

$$\sum_{l} \binom{k}{l}^2 w^l = \sum_{l} \binom{k}{l} \operatorname*{res}_{z} \frac{(1+z)^k}{z^{l+1}} w^l = \operatorname*{res}_{z} \frac{(1+z)^k}{z} \left(1 + \frac{w}{z}\right)^k,$$

we obtain,

$$\begin{aligned} S(N) &= \sum_{k} \binom{N}{k}^2 w^k \operatorname*{res}_{z} \frac{(1+z)^k (w+z)^k}{z^{k+1}} \\ &= \langle z^0 \rangle \sum_{k} \binom{N}{k}^2 \left(\frac{w(1+z)(w+z)}{z} \right)^k. \end{aligned}$$

The observation,

$$\frac{w(1+z)(w+z)}{z} - 1 = \frac{(1-wz)^2}{wz} \quad \text{if} \quad w = e^{2\pi i/3},$$

suggests to invoke binomial expansion together with the fact that $\langle z^0 \rangle g(cz) = \langle z^0 \rangle g(z)$:

$$\begin{aligned} S(N) &= \langle z^0 \rangle \sum_{k} \binom{N}{k}^2 \left(1 + \frac{(1-wz)^2}{wz}\right)^k \\ &= \langle z^0 \rangle \sum_{k,j} \binom{N}{k}^2 \binom{k}{j} \left(\frac{(1-wz)^2}{wz} \right)^j \\ &= \sum_{k,j} \binom{N}{k}^2 \binom{k}{j} \langle z^0 \rangle \left(\frac{(1-z)^2}{z} \right)^j \\ &= \sum_{k,j} \binom{N}{k} \binom{N-j}{N-k} \binom{N}{j} \binom{2j}{j} (-1)^j. \end{aligned}$$

For the last equality we applied

$$\binom{N}{k}\binom{k}{j} = \binom{N-j}{N-k}\binom{N}{j}.$$

Finally, invoking Vandermonde's formula on the sum over k reduces $S(N)$ to the single sum

$$S(N)=\sum_j \binom{2N-j}{N}\binom{N}{j}\binom{2j}{j}(-1)^j. \tag{7.5}$$

7.2.2 *Simplifying the single sum*

7.2.2.1 Using computer algebra

Taking the single sum in (7.5) as input, any implementation of Zeilberger's algorithm [14] reveals that

$$(N+3)^2S(N+3)-4(4N+3)(4N+9)S(N)=0 \qquad (N\geq 0). \tag{7.6}$$

This recurrence together with the initial values $S(0)=1$, $S(1)=0$, and $S(2)=0$ proves the evaluation stated above.

7.2.2.2 Using classical hypergeometric machinery

We find it instructive to present an alternative evaluation of the single sum in (7.5), namely with classical hypergeometric methods. See [5] for an introduction to basic notions and terminology or, for a more detailed account, the monograph [1].

First of all, we rewrite (7.5) in hypergeometric series notation:

$$S(N)=\binom{2N}{N}\,{}_3F_2\left(\begin{matrix}-N,-N,\frac{1}{2}\\ -2N,1\end{matrix};4\right). \tag{7.7}$$

Next, consider an important but less known cubic transformation of W. N. Bailey [2, (4.06)]:

$$\begin{aligned}{}_3F_2\left(\begin{matrix}3a,b,3a-b+\frac{1}{2}\\ 2b,6a-2b+1\end{matrix};4x\right)&=(1-x)^{-3a}\\ &\times{}_3F_2\left(\begin{matrix}a,a+\frac{1}{3},a+\frac{2}{3}\\ b+\frac{1}{2},3a-b+1\end{matrix};\frac{27x^2}{4(1-x)^3}\right).\end{aligned} \tag{7.8}$$

For complex numbers a, b, and x the identity holds if $|x|\leq 1/4$; in this case both ${}_3F_2$ series converge absolutely. If $a=-N/3$ and $b=-N$ for a positive integer N, then identity (7.8) holds for all x. Namely, in this case both ${}_3F_2$ series terminate, and both sides of (7.8) turn into polynomials in x of degree N. In particular, note that the right hand side of (7.8) can be viewed as a polynomial expanded in terms of polynomials of the form $x^{2k}(1-x)^{N-3k}$ where $0\leq k\leq N/3$.

Finally observe that $S(N)$ as given in (7.7) is nothing but $\binom{2N}{N}$ times the left hand side of (7.8) with the choice $a = -N/3$, $b = -N$, and $x = 1$. According to (7.8) this equals 0 if N is not divisible by 3. If $N = 3n$ then $S(3n)$ equals $\binom{6n}{3n}$ times the coefficient of $x^{2k}(1-x)^{3n-3k}$ with $k = n$ of the polynomial on the right hand side of (7.8), which gives

$$\binom{6n}{3n} \text{ times } \frac{(-n)_n(-n+\frac{1}{3})_n(-n+\frac{2}{3})_n}{(-3n+\frac{1}{2})_n(1)_n}\frac{\left(\frac{27}{4}\right)^n}{n!}.$$

This can be rewritten as $(4n)!/(n!n!(2n)!)$, in accordance with (7.1).

7.3 MultiSum in Action

Instead of first reducing the original problem to a single sum, one could apply computer algebra *directly* to the given double sum to obtain the recurrence (7.6). We illustrate this approach by using Wegschaider's Mathematica package `MultiSum` [12], which is based on WZ summation [13]. We initialize by loading the package

```
In[1]:= <<MultiSum.m
```

The first step is to compute a "certificate recurrence" for the summand:

```
In[2]:=FindRecurrence[
  Binomial[N,k]^2 Binomial[k,l]^2 w^(k + l),{N},{k,l}]
Out[2]= <certificate recurrence>
```

Without going into details it suffices to note that the output `<certificate recurrence>` is of the form

$$\sum_{(r,s,t)\in I} p_{r,s,t}(N,w)F(N-r,k-s,l-t) = \Delta_k Q_w^{(1)}(N,k,l) + \Delta_l Q_w^{(2)}(N,k,l). \quad (7.9)$$

Here $F(N,k,l)$ denotes the summand $\binom{N}{k}^2\binom{k}{l}^2 w^{l+k}$, I denotes some index set for the integer shifts, the $p_{r,s,t}(N,w)$ are polynomials in N and w (free of the summation variables k and l), and the $Q_w^{(i)}(N,k,l)$ are of the form as the left hand side of (7.9) with different coefficient polynomials and index sets. The symbol Δ_m denotes the difference operator, i.e., $\Delta_m g(m) = g(m+1) - g(m)$. Note that for fixed N the support of $F(N,k,l)$ with respect to integers k and l is finite.

Due to the finite support property we can sum both sides of (7.9) over all integers k and l. This produces a recurrence for $S(N) = \sum_{k,l} F(N,k,l)$ which is executed by the command:

```
In[3]:=ShiftRecurrence[SumCertificate[%],{N,1}]
Out[3]=<w-recurrence>
```

The output <w-recurrence> is of the form:

$$p_{-1}(N,w)\ \mathrm{SUM}[N-1]+p_0(N,w)\ \mathrm{SUM}[N]+\cdots+p_3(N,w)\mathrm{SUM}[N+3]=0$$

with $\mathrm{SUM}[N]=S(n)$ and where the $p_i(N,w)$ are polynomials in N and w.

Finally the recurrence (7.6) is obtained by the substitution $w=e^{2\pi i/3}$:

```
In[4]:=FullSimplify[% /.w->Exp[2PiI/3]]

Out[4]:={(4N+1)(4N+5)
         (4(4N+3)(4N+9)SUM[N]-(N+3)^2 SUM[N+3])==0}
```

7.4 GeneratingFunctions in Action

A completely different approach to Henrici's friendly monster (7.2) is via the use of generating functions.

Considering the special case $c=1$ of $f(c,x)$ in (7.3), we define

$$F(x) \doteqdot f(1,x)=\sum_{j=0}^{\infty}\frac{x^j}{(j!)^2}.$$

Recalling $S(N)=\sum_{k+l+m=N}\left(\frac{N!}{k!l!m!}\right)^2 w^{l-m}$, it is easy to verify that

$$F(x)\ F(wx)\ F(w^2x)=\sum_{N=0}^{\infty}S(N)\frac{x^N}{(N!)^2}. \qquad (7.10)$$

Our strategy to derive the desired closed form for $S(N)$ will make use of D-finite (also called: holonomic) closure properties. More precisely: Starting with a differential equation for $F(ux)$, we first will derive a differential equation for the left hand side of (7.10), which in the next step will be converted to recurrences for $S(N)/(N!)^2$ and $S(N)$, respectively. All these steps will be carried out automatically by using Mallinger's Mathematica package `GeneratingFunctions` [9]. In the Maple system analogous procedures are available; see the pioneering work [10]. For more detailed theoretical background information consult [11, Sect. 6.4].

We initialize by loading the package

```
In[1]:=<<GeneratingFunctions.m
```

First we transform the recurrence for the coefficient sequence $a[j]=u^j/(j!)^2$ into a differential equation for $F(ux)$:

In[2]:=DE[u_]:=RE2DE[$(j+1)^2 a[j+1]-u\,a[j]==0$, $a[j]$, $g[x]$]

In[3]:=DE[u]

Out[3]:= $-ug[x]+g'[x]+xg''[x]=0$

Then we derive a differential equation for $g(x) \doteqdot F(x)F(wx)F(w^2x)$:

```
In[4]:=DiffEq=FullSimplify[
DECauchy[DECauchy[DE[1],DE[w],g[x]],DE[w^2],g[x]]/. w → e^(2πi/3)]
```

Out[4]= $108\,g[x]+256\,x\,g'[x]==54\,g^{(3)}[x]+x(330\,g^{(4)}[x]+x \times(-64\,g''[x]+393\,g^{(5)}[x]+153\,x\,g^{(6)}[x]+22\,x^2\,g^{(7)}[x]+x^3\,g^{(8)}[x]))$

The differential equation `DiffEq` for $g(x) \doteqdot F(x)F(wx)F(w^2x)$ is transformed into a recurrence for the coefficients $c[N] = S(N)/(N!)^2$ in the Taylor expansion of $g(x) = \sum_{N\geq 0} c[N]x^N$:

In[5]:=crec=DE2RE[DiffEq, $g[x], c[N]$]

Out[5]= $4(4N+3)(4N+9)c[N]-(N+1)^2(N+2)^2(N+4)^4c[N+3]==0$

Finally, to obtain the desired recurrence for $S(N)$ we multiply $c[N]$ with $(N!)^2$; again this operation is carried out on the recurrence representations:

In[6]:=REHadamard[crec, $c[N+1]==(N+1)^2c[N]$, $c[N]$]

Out[6]= $-4(4N+3)(4N+9)c[N]+(N+3)^3c[N+3]==0$

This way we again arrived at recurrence (7.6).

Remarks. (a) The full form (7.2) of the friendly monster can be proven automatically with executing exactly the same steps!
(b) From Mallinger's package we used the procedure calls: `RE2DE`, `DECauchy`, `DE2RE`, and `REHadamard`. Full descriptions of their functionality can be found in [9].
(c) It is instructive to compare the steps of the computer derivation in this Section to Henrici's original proof in [8] which also uses differential equations. In his concluding remark [8, p. 1518], Henrici seems to anticipate future computer developments: "The method used in this work has the advantage of not requiring to be known in advance. Although the algebraic manipulations, if done by hand, soon become unmanagable, the method can be used in principle whenever the irreducible terms in the derivatives of a product satisfy a recurrence relation of the general form [as specified in the paper]."

7.5 Conclusion

In Section 7.2.1 we have seen how Egorychev's method can be applied to reduce a multiple sum to a single one. The single sum then was reduced by Zeilberger's algorithm. We have shown in Section 7.3 that computer algebra already can be applied to the originally given multiple sum in the version of (7.4). Namely, using Wegschaider's package `MultiSum`, we derived recurrences which led to the desired simplification. Despite not discussed explicitly in the text, we want to note that both of these computer algebra applications are such that the programs also produce proofs, respectively proof certificates, that can be used for (human) verification of the computer output - independently from the particular steps of the algorithms.

Moreover, three aspects of Egorychev's method should be emphasized. First, in contrast to the `MultiSum` approach, it sheds additional light on the structure of the problem. Namely, in Section 7.2.2.2 we have seen that it reduced the original problem to a special instance of Bailey's cubic transformation (7.8). Second, it is well known that computer algebra algorithms for simplifying multiple sums like `MultiSum` in various applications still struggle with efficiency. In such situations, a method like Egorychev's can be applied as a preprocessing step to reduce complexity. Third, for these reasons we suggest future investigations whether certain aspects of Egorychev's approach could be supported with symbolic computation. In the context of MacMahon's partition analysis [16] some developments have been done in this direction; see [15] and in particular the algorithm by Han [6].

Finally the algorithmic approach used in Section 7.4 gives rise to the following remarks. It produced no proof certificates, but replaces Henrici's cumbersome hand manipulations by automatic procedures. In addition, it provides insight into the structure of the problem, and also allows to generalize; see the Remarks above. There we pointed out that in exactly the same fashion one can derive a proof of the full monster identity (7.2). We have not investigated whether this applies also to the derivation carried out in Section 7.2. More precisely: Can the Egorychev approach as used in Section 7.2 be modified to prove the full monster identity?

Acknowledgement. I want to thank Professor Johann Cigler who during my student days at the University of Vienna has introduced me and my colleagues to Egorychev's book [3] in the frame of a highly inspiring seminar.

References

1. G. E. Andrews, R. Askey, and R. Roy, *Special Functions*, Cambridge University Press, 1999.
2. W. N. Bailey, *Products of generalized hypergeometric series*, Proc. London Math. Soc. **28** (1928), 242–254.
3. G. P. Egorychev, *Integral Representation and the Computation of Combinatorial Sums*, Translations of Mathematical Monographs, Vol. 59, Amer. Math. Soc., Providence, RI, 1984. (Translation of: Integral'noe predstavlenie i vychislenie kombinatornykh summ.)
4. R. L. Graham, D. E. Knuth, and O. Patashnik, *Concrete Mathematics*, 1st edition, Addison-Wesley, 1989.
5. R. L. Graham, D. E. Knuth, and O. Patashnik, *Concrete Mathematics*, 2nd edition, Addison-Wesley, 1994.
6. G. N. Han, *A general algorithm for the MacMahon Operator*, Ann. of Comb. **7** (2003), 467–480.
7. P. Henrici, *De Branges' proof of the Bieberbach conjecture: a view from computational analysis*, Sitzungsber. der Berliner Math. Ges. (1987), 105–121.
8. P. Henrici, *A triple product theorem for hypergeometric series*, SIAM J. Math. Anal. **18** (1987), 1513–1518.
9. C. Mallinger, *Algorithmic Manipulations and Transformations of Univariate Holonomic Functions and Sequences*, Master's Thesis (Diplomarbeit), RISC, Johannes Kepler University Linz, 1996. Available at: http://www.risc.uni-linz.ac.at/research/combinat.

10. B. Salvy and P. Zimmermann, *Gfun: a package for the manipulation of generating and holonomic functions in one variable*, ACM Trans. Math. Software **20** (1994), 163–177.
11. R. P. Stanley, *Enumerative Combinatorics Vol. 2*, Cambridge University Press, 1999.
12. K. Wegschaider, *Computer Generated Proofs of Binomial Multi-Sum Identities*, Master's Thesis (Diplomarbeit), RISC, Johannes Kepler University Linz, 1997. Available at: http://www.risc.uni-linz.ac.at/research/combinat.
13. H. S. Wilf and D. Zeilberger, *An algorithmic proof theory for hypergeometric (ordinary and "q") multisum/integral identities*, Inventiones Math. **108** (1992), 575–633.
14. D. Zeilberger, *A fast algorithm for proving terminating hypergeometric identities*, Discrete Math. **80** (1982), 207–211.
15. G. E. Andrews and P. Paule, *MacMahon's Partition Analysis IV: Hypergeometric multisums*, Séminaire Lotharingien de Combinatoire **42** (1998), Paper B42i. (Also in: *The Andrews Festschrift: Seventeen Papers on Classical Number Theory and Combinatorics*, D. Foata and G.-N. Han, eds., pp. 189–208. Berlin, Springer, 2001.)
16. G. E. Andrews, P. Paule and A. Riese, *MacMahon's Partition Analysis VI: A new reduction algorithm*, Annals Combin. **5** (2002), 251–270.

Chapter 8
The Automatic Central Limit Theorems Generator (and Much More!)

Doron Zeilberger

Dedicated to Georgy Petrovich EGORYCHEV on his 70th birthday

Why I hate the Continuous and Love the Discrete

I have always loved the discrete and hated the continuous. Perhaps it was the trauma of having to go through the usual curriculum of "rigorous", Cauchy-Weierstrass-style, real calculus, where one has all those tedious, pedantic and *utterly boring*, $\varepsilon-\delta$ proofs. The meager (obvious) conclusions hardly justify the huge mental efforts! Complex Analysis was a different story. Even though officially "continuous", it has the feel of discrete math, and one can "cheat" and consider power series as formal power series, and I really loved it.

Georgy P. Egorychev: A Bridge-Builder between the Discrete and the Continuous

Eight years after I finished my doctorate, I came across Egorychev's fascinating modern classic [2], about using the methods of complex analysis to evaluate (discrete) combinatorial sums. That was a pioneering ecumenical work, that influenced me greatly. Its content, of course, but especially its *spirit* and *philosophy*.

The Discrete vs. The Continuous: A Two-Way Street

Egorychev went from the discrete to the continuous. But the *bridge* that he helped build can be transversed **both** ways. With the advent of so-called **Wilf-Zeilberger** (WZ) theory (see [8]) one can indeed go both ways. Sometimes the discrete is easier to handle, and sometimes the continuous. But nothing is really *continuous*. There

---

Doron Zeilberger
Department of Mathematics, Rutgers University (New Brunswick), Hill Center-Busch Campus, 110 Frelinghuysen Rd., Piscataway, NJ 08854-8019, USA, e-mail: zeilberg@math.rutgers.edu
Accompanied by Maple packages `CLT` and `AsymptoticMoments` downloadable from the webpage http://www.math.rutgers.edu/~zeilberg/mamarim/mamarimhtml/georgy.html, where one can also find some sample input and output files. Supported in part by the NSF.

I.S. Kotsireas, E.V. Zima (eds.), *Advances in Combinatorial Mathematics*,
DOI 10.1007/978-3-642-03562-3_8, © Springer-Verlag Berlin Heidelberg 2009

is only the *discrete* and the "*continuous*", the quotation-marks indicating that it is really discrete in disguise, and, on a fundamental level, continuous mathematics is just a degenerate case of the discrete, as I have already preached in [9].
Initially, I was hoping to write something about interfacing Egorychev's brilliant approach with WZ theory, but meanwhile I got distracted by another project, that also has the **discrete-continuous** theme, namely for automatically deriving *limit laws* in probability theory, and decided to make this my *tribute* to Georgy Egorychev's 70th birthday.

Probability Limit Laws

One of the central themes of modern probability theory are *limit laws*, the most celebrated one being the Central Limit Theorem, that roughly says that if you repeat the same experiment many times, and the "atomic" experiment can have an arbitrary probability distribution (with finite variance), then in the limit, after one "centralizes" and "normalizes" (divides by the so-called standard deviation) one gets the (continuous) *Standard Normal Distribution*:

$$Pr(a \leq X \leq b) = \frac{1}{\sqrt{2\pi}} \int_a^b e^{-\frac{x^2}{2}}\, dx.$$

The *iconic* example of a discrete probability distribution is the random variable "number of Heads" upon tossing a (loaded) coin n times, whose probability distribution is given by the **Binomial Distribution**, usually denoted by $\mathbf{B}(n,p)$. It describes the experiment of tossing a coin n times with the probability of Heads being p. The "sample space" is the set of all 2^n outcomes $\{H,T\}^n$, and the probability of an "atomic" event is $p^{NumberOfHeads}(1-p)^{NumberOfTails}$, and hence the probability of the "compound event", *NumberOfHeads=k*, is $\binom{n}{k} p^k (1-p)^{n-k}$. If we call this random variable X_n, then its mean (see below) is np and its variance (also see below) is $\sigma^2 := np(1-p)$. Introducing the *centralized and normalized* random variable

$$Z_n := \frac{X_n - np}{\sqrt{np(1-p)}},$$

The "original" (De Moivre-Laplace) Central Limit Theorem asserts that

$$Z_n \to \mathcal{N},$$

where $\mathcal{N}$ is the Standard Normal Distribution.
More generally, quoting from Feller ([3], p. 244):
Central Limit Theorem. Let $\{X_k\}$ be a sequence of mutually independent random variables with a common distribution. Suppose that $\mu := \mathbf{E}[X_k]$ and $\sigma^2 := Var[X_k]$ exist and let $\mathbf{S}_n = X_1 + \cdots + X_n$. Then for every fixed β,

$$\mathbf{P}\left\{\frac{\mathbf{S}_n - n\mu}{\sigma\sqrt{n}} < \beta\right\} \to \mathcal{N}(\beta),$$

where $\mathcal{N}(x)$ is the normal distribution defined above.
There are many extensions and generalizations. In this article we will present yet another extension, but in a completely different direction, and because of the heavy use of computers, we are pretty sure that these are new results.

A Quick Review of Discrete Probability Distributions

The most basic scenario is that we have a *finite* set S, called the *sample space*, consisting of *atomic events*, and each $s \in S$ has a certain probability (a number in $[0,1]$) attached to it, where, of course, $\sum_{s\in S} p_s = 1$.
We also have a *random variable* $X : S \to R$, where R is a finite set of real numbers (often, but not always, of integers), and one is interested in its *probability* distribution $Pr(\{s \in S | X(s) = r\})$. A convenient way to encode it is via its, *probability generating function*,

$$f(t) := \sum_{r\in R} Pr(X(s) = r) t^r,$$

that is easily seen to be equal to the *weighted counting* of the set S

$$\sum_{s\in S} p_s t^{X(s)}.$$

The most important number associated to a random variable is its *expectation*

$$\mu = E[X] := \sum_{s\in S} p_s X(s).$$

This is also called the *first* moment. Analogously, the *higher moments* (about the mean) are defined by

$$m_r(X) := \sum_{s\in S} p_s (X(s) - \mu)^r.$$

It follows from "general nonsense" that, under some mild conditions (that are always satisfied for finite sets), the moments completely determine the probability distribution (even in the general, "infinite", case), and the probability distribution can be gotten by inverse-Fourier-Transforming the **moment (exponential) generating function** $\sum_r m_r (it)^r / r! = E[exp(itX)]$.
Another set of moments, easier to work with, are the *factorial moments*

$$f_r(X) := \sum_{s\in S} p_s (X(s) - \mu)^{(r)},$$

where $X^{(r)}$ is the *falling factorial*:

$$X^{(r)} := X(X-1)(X-2)\dots(X-r+1).$$

It turns out to be easier (see below) to compute the factorial moments, but once these are known, one can get the ordinary moments, thanks to the **connection formula** (e.g. [4], p. 250):

$$X^r = \sum_{k=1}^{r} S(r,k) X^{(k)},$$

where $S(r,k)$ are the *Stirling Numbers of the Second kind*, that may be defined by the recurrence ([4], p. 250):

$$S(r,k) = kS(r-1,k) + S(r-1,k-1), \qquad (StirlingRecurrence)$$

subject to the initial condition $S(1,k) = 1$ if $k = 1$ and $S(1,k) = 0$ otherwise.
It follows that the moments can be computed in terms of the factorial moments:

$$m_r = \sum_{k=1}^{r} S(r,k) f_r.$$

Computing Moments

Suppose that we have the probability generating function $f(t)$. We can find its *mean*, μ, by differentiating with respect to t, and plugging-in $t = 1$:

$$\mu = f'(1).$$

Immediately we can find the probability generating function of the centralized random variable $X_C(s) := X(s) - \mu$. It is simply

$$\frac{f(t)}{t^{\mu}}.$$

From now, let's assume that all our random variables have mean 0, in other words, assume that we have already done this centralization, and let's rename it $f(t)$. Using the new, adjusted, $f(t)$, we can easily find the factorial moments, by taking successive derivatives, and substituting $t = 1$ at the end:

$$f_r = \left. \frac{d^r f(t)}{dt^r} \right|_{t=1}.$$

Alternatively, we can consider $f(1+z)$ and do a Maclaurin expansion around $z = 0$:

$$f(1+z) = \sum_{r=0}^{\infty} f_r \frac{z^r}{r!}.$$

Repeating It n Times

So far what we said is true in general. A frequently occurring situation is when we repeat something n times, like tossing a coin, or rolling a die, and we are interested in the sum of the outcomes. In that case, we have a sequence of random variables whose probability generating function is

$$F(t)^n,$$

where $F(t)$ is the probability generating function for the single event. For example, for tossing a single coin, where the random variable is "number of Heads", and the probability of a Head is p, we have

$$F(t) = \frac{pt + (1-p)}{t^p} = pt^{1-p} + (1-p)t^{-p},$$

and for rolling a loaded (cubic) die, with its probabilities of landing on $1,2,3,4,5,6$ being $p_1, p_2, p_3, p_4, p_5, p_6$ respectively, (where of course $p_1 + \cdots + p_6 = 1$), is

$$F(t) = \frac{\sum_{i=1}^{6} p_i t^i}{t^{\mu}}, \quad \text{where} \quad \mu := \sum_{i=1}^{6} i p_i,$$

etc.

To get the first R factorial moments, for any specific, desired R, we simply find the first $R+1$ terms in the Taylor expansion of $F(t)^n$, at $t = 1$, that Maple can easily do symbolically, getting *explicit* polynomial expressions, in n, for the r-th factorial moment, for each specific, numeric, r. What it *can't* do is find the general expression for *symbolic* r (as well as n, of course).

An even more efficient way to crank-out explicit polynomial expressions for the factorial moments, is to, *once and for all*, crank out sufficiently many coefficients of $F(1+z)$ itself (equivalently find sufficiently many factorial moments of the "atomic" experiment), let's call them F_i, where, of course, $F_0 = 1$ and $F_1 = 0$.

$$F(1+z) = 1 + \sum_{r=2}^{\infty} \frac{F_r}{r!} z^r,$$

and then use the obvious fact that

$$F(1+z)^{n+1} = F(1+z)^n \cdot F(1+z)$$

that entails:

$$1 + \sum_{r=2}^{\infty} \frac{f_r(n+1)}{r!} z^r = \left(1 + \sum_{r=2}^{\infty} \frac{f_r(n)}{r!} z^r\right)\left(1 + \sum_{r=2}^{\infty} \frac{F_r}{r!} z^r\right).$$

Rearranging, and comparing coefficient of z^r, we have the following *recurrence*

$$f_r(n+1) - f_r(n) = \sum_{s=2}^{r} \binom{r}{s} F_s f_{r-s}(n), \qquad (Recurrence)$$

Since obviously $f_r(0) = 0$, this uniquely determines $f_r(n)$ as the indefinite sum of the right side, and it immediately follows by induction that the even factorial moments $f_{2r}(n)$ are polynomials of degree r, and the odd factorial moments $f_{2r+1}(n)$ are also polynomials of degree r. (Of course $f_1(n) = 0$).

Asymptotic Factorial Moments

There is no way that we can get an *explicit*, symbolic, expression, in **both** n and r for the general factorial moments $f_{2r}(n), f_{2r+1}(n)$. But, thanks to the miracle of computers, we can get **explicit** expressions for their s-leading terms *for any desired* s.
Either "cheating" and using our knowledge that the normalized even factorial moments $f_{2r}(n)/f_2(n)^r$ should tend to the even moments $(2r)!/(2^r r!)$ of the Standard Normal Distribution, and the normalized odd factorial moments $f_{2r+1}(n)/f_2(n)^{r+1/2}$ should tend to the odd moments (0) of the Standard Normal Distribution, but better still, doing it *ab initio*, by staring at the leading terms and making the obvious conjectures, we can write:

$$f_{2r}(n) = f_2(n)^r \frac{(2r)!}{2^r r!}\left[\left(1+\sum_{i=1}^{s}\frac{A_i(r)}{n^i}\right)+O\left(\frac{1}{n^{s+1}}\right)\right],$$

and, analogously

$$f_{2r+1}(n) = f_2(n)^r \frac{(2r)!}{2^r r!}\left[\left(\sum_{i=0}^{s}\frac{B_i(r)}{n^i}\right)+O\left(\frac{1}{n^{s+1}}\right)\right].$$

(Note that $f_2 = nF_2$).
Substituting this *ansatz* into (*Recurrence*), it emerges that the $A_i(r)$'s and $B_i(r)$'s are certain polynomials in r. Rather than untangle the complicated implied recurrences for them, we empirically, in turn, for $i = 0, 1, 2, \ldots$, crank-out $A_i(r), B_i(r)$ for sufficiently many numeric r and then "fit" appropriate polynomials, using *undetermined coefficients* in the context of the *polynomial ansatz* (see [10]). Once we have the *conjectured* explicit expressions, for the asymptotic expansion up to our desired order $(1/n^s)$, we can, *a posteriori*, prove them *rigorously* by verifying (*Recurrence*) to that desired order.
The Central Limit Theorem only asserts that the normalized r-th moments converge to the moments of the Standard Normal Distributions, i.e. the case $s = 0$. So in particular, our computer reproved the Central Limit Theorem, but with a *vengeance*, it gave us the first s terms in the asymptotics, where s is as big as we wish (of course the higher the s, the longer that it would take).

What about the ordinary moments?

From

$$m_r = \sum_{k=1}^{r} S(r,k) f_r,$$

we get:

$$m_r(n) = \sum_{k=0}^{s} S(r, r-k) f_{r-k}(n) + O\left(\frac{1}{n^{s+1}}\right).$$

Define

$$S_k(r) := S(r, r-k).$$

It is easy to see that $S_k(r)$ are polynomials in r of degree $2k$. Indeed the defining recurrence (*StirlingRecurrence*) transcribes to:

$$S_k(r) - S_k(r-1) = (r-k)S_{k-1}(r-1),$$

from which Maple can easily compute, recursively, as many of the $S_k(r)$ as needed, starting at the obvious initial condition $S_0(r) = 1$, and taking the indefinite sum, with respect to r, of the already known right hand side.
So to get the up-to-order-s asymptotics for the ordinary moment $m_r(n)$, Maple simply computes, *all by itself,*

$$m_r(n) = \sum_{k=0}^{s} S_k(r) f_{r-k}(n) + O\left(\frac{1}{n^{s+1}}\right),$$

using the already computed expressions (in *symbolic* r and n) for f_{2r} and f_{2r+1} obtained above (up to the desired order s). Of course, we would have to treat the even moments, m_{2r}, and the odd moments m_{2r+1} separately, and they obviously have different expressions, but the computer does not mind.

Repeating n times a Generic Probability Distribution

The above discussion applies equally to repeating a general probability distribution, given by its ordinary moments $M_1 = 0, M_2 = 1, M_3, M_4 \ldots$. One first finds the factorial moments (now using the Stirling numbers of the *first* kind), and using the above formula, one can get the asymptotics of the moments of the "repeated" n-times random variable, to any desired order s, of the $2r$-th and $(2r+1)$-th moments for the normalized sum of n repetitions. For example, the first term is:

$$1 + \frac{(-1+r)\,r\left(2\,rM_3{}^2 + 3M_4 - 9 - 4M_3{}^2\right)}{18n} + O\left(\frac{1}{n^2}\right).$$

More terms are available at the webpage of this article.
This leads us to the following interesting observation, that, once made, should be provable using moment generating functions.
Refined Central Limit Theorem. Let $\{X_k\}$ be a sequence of mutually independent random variables with a common distribution. Suppose that $\mu := \mathbf{E}[X_k] = 0$ and $\sigma^2 := E[X^2] = 1$, and all the first $2s$ moments, $M_1 = 0, M_2 = 1, M_3, M_4, \ldots, M_{2s}$, are finite. Let $\mathbf{S}_n = X_1 + \cdots + X_n$, and let $m_{2r}(n)$ be the $2r$-th moment of $\mathbf{S}_n$. Then for even s,

$$m_{2r}(n) = (2r)!/(2^r r!)(1 + O(1/n^s))$$

if the first $2s$ moments of X are the same as the first $2s$ moments of the Standard Normal Distribution (namely: 0,1,0,3,0,15,0,105, ...) .

Limit Laws for Sequences of Discrete Probability Distributions

The Central Limit Theorem talks about the limit of a family of discrete probability distributions, whose probability generating functions are given by the extremely simple

$$P_n(t) := F(t)^n,$$

that satisfy a first-order recurrence with *constant*, (in n) coefficients

$$P_{n+1}(t) = F(t)P_n(t).$$

Many natural families of discrete probability distributions, especially those arising from generating functions in combinatorial enumeration ("q-counting"), satisfy a more general kind of first-order recurrence:

$$P_{n+1}(t) = F(n,t,t^n)P_n(t),$$

where $F(n,t,t^n)$ is a certain explicit rational function of n,t,t^n.
For example (switching to the letter q to respect combinatorial tradition), consider the set of permutations on n elements, under the "mahonian" statistics, whose *counting* generating function is:

$$\prod_{i=1}^{n} \frac{1-q^i}{1-q}.$$

The expectation is, of course, $n(n-1)/4$, so dividing by $n!q^{n(n-1)/4}$, we get that the probability generating function for the random variable "number of inversions" is:

$$P_n(q) = \prod_{i=1}^{n} \frac{q^{-i/2}-q^{i/2}}{i(q^{-1/2}-q^{1/2})}.$$

So, in this case,

$$F(n,q,q^n) = \frac{q^{-(n+1)/2}-q^{(n+1)/2}}{(n+1)(q^{-1/2}-q^{1/2})}.$$

See [3] (sec. X.6),[5, 6, 7] for other approaches for proving Asymptotic Normality.
Another example is the q-Catalan distribution, whose asymptotic normality has been recently proved by Chen, Wang, and Wang [1], who also proved more general results.
The discussion in the previous section goes almost verbatim to such more general families of discrete probability distributions, except that now we can no longer (always) find the first factorial moments directly. Now we *must* use the generalization of $(Recurrence)$.
Instead of

$$F(1+z) = 1 + \sum_{r=2}^{\infty} \frac{F_r}{r!} z^r,$$

we now have:

$$F(n, 1+z, (1+z)^n) = 1 + \sum_{r=2}^{\infty} \frac{F_r(n)}{r!} z^r,$$

where now the $F_r(n)$ are polynomials of n, and no longer have an interpretation as factorial moments for an "atomic event". They are just the Maclaurin coefficients of this more general object. The analog of (*Recurrence*) reads:

$$f_r(n+1) - f_r(n) = \sum_{s=2}^{r} \binom{r}{s} F_s(n) f_{r-s}(n). \qquad (GeneralRecurrence)$$

The same empirical approach as before still applies. We normalize, and guess explicit expressions for the coefficients $A_i(r), B_i(r)$ in the normalized factorial moments, that are then rigorously proved *a posteriori*. Once explicit asymptotic expressions for the even and odd factorial moments have been derived and proved, one uses the Stirling polynomials in order to deduce explicit expressions for the even and odd (usual) moments (about the mean), in particular proving *asymptotic normality*, but with precise asymptotic expansion, to any desired order, of the general moments.

Accompanying Maple Packages

This article is accompanied by two Maple packages: `AsymptoticMoments`, and `CLT`. Most of the procedures in `CLT` are subsumed by the more general procedures of `AsymptoticMoments`, but the former has some extra features. These packages can be downloaded from the webpage: http://www.math.rutgers.edu/~zeilberg, where there is ample sample input and output.

In particular, `AsymptoticMoments` is applied to the above-mentioned cases of the mahonian and q-Catalan distribution, thereby sharpening them, by not only proving asymptotic normality, but presenting a more detailed asymptotics for the moments. We also post numerous other examples, for example, plane partitions whose 3D Ferrers diagrams are bounded in a box of any given, (numeric) height.

Let us cite the simplest output. If you toss a fair coin n times, then the $2r$-th moment is $(n/4)^r (2r)!/(2^r r!)$ times

$$1 - 1/3 \frac{(r-1)r}{n} + \frac{1}{90} \frac{r(r-1)(r-2)(5r+1)}{n^2} -$$
$$\frac{1}{5670} \frac{(r-1)(r-2)(r-3)(35r^2+21r-32)r}{n^3} + O\left(\frac{1}{n^4}\right).$$

Conclusion: Why is this interesting?

Locally, it is interesting for its own sake, but *globally* it is interesting since it presents a beautiful example how probability theory would have been very different, had the computer been available three hundred years ago. Using *symbol-crunching* the computer can derive deep theorems, and largely obviates all the human attempts at a "rigorous" foundation of continuous probability, using measure theory and Kolmogorov's "axiomatic" approach. The passage from the discrete to the continuous becomes much more concrete and down-to-earth, and it is apparent that Discrete Math rules, and Continuous Math is indeed a degenerate case. For other examples of probability computerized-redux, see [11].

Future Work

This is just the tip of an iceberg. One should be able to consider much larger families of discrete probability distributions, not just those given by first-order recurrences. Also *joint distributions*, and *multivariate* limit laws should be amenable to the present approach. For example, proving the joint asymptotic normality of the number of inversions and the *major index* on the set of permutations on $\{1,2,\ldots,n\}$, using the more complicated recurrences derived in [12].

References

1. W.Y.C. Chen, C.J. Wang, and L.X.W. Wang, *The Limiting Distribution of the coefficients of the q-Catalan Numbers*, Proc. Amer. Math. Soc. **136** (2008), 3759-3767.
2. G.P. Egorychev, "*Integral Representation and the Computation of Combinatorial Sums*", (translated from the Russian), Amer. Math. Soc., Providence, 1984.
3. William Feller, "*An Introduction to Probability Theory and Its Application*", volume 1, three editions. John Wiley and sons. First edition: 1950. Second edition: 1957. Third edition: 1968.
4. R. Graham, D.E. Knuth, and O. Patashnik, *Concrete Mathematics: A Foundation for Computer Science*, Addison Wesley, Reading, 1989.
5. G. Louchard and H. Prodinger, *The number of inversions in permutations: a saddle point approach*, J. Integer Seq. **6** (2003), A03.2.8.
6. B.H. Margolius, *Permutations with inversions*, J. Integer Seq. **4** (2001), A01.2.4.
7. V.N. Sachkov, "*Probabilistic Methods in Combinatorial Analysis*", Cambridge University Press, New York, 1997.
8. H. Wilf and D. Zeilberger, *An algorithmic proof theory for hypergeometric (ordinary and "q") multisum/integral identities*, Invent. Math. **108** (1992), 575-633.
9. Doron Zeilberger, *"Real" Analysis is a Degenerate Case of Discrete Analysis*. In: "New Progress in Difference Equations" (Proc. ICDEA 2001), edited by Bernd Aulbach, Saber Elaydi, and Gerry Ladas, Taylor & Francis, London, 2004.
10. Doron Zeilberger, *An Enquiry Concerning Human (and Computer!) [Mathematical] Understanding*. In: C.S. Calude, ed., "Randomness & Complexity, from Leibniz to Chaitin", World Scientific, Singapore, Oct. 2007.
11. Doron Zeilberger, *Fully AUTOMATED Computerized Redux of Feller's (v.1) Ch. III (and Much More!)*, Personal Jouranl of Ekhad and Zeilberger (Nov. 14, 2006), `http://www.math.rutgers.edu/~zeilberg/pj.html`.
12. Doron Zeilberger, *A Lattice Walk Approach to the "inv" and "maj" q-counting of Multiset Permutations*, J. Math. Anal. Applications **74** (1980), 192-199.

MIX
Papier aus verantwortungsvollen Quellen
Paper from responsible sources
FSC® C105338

If you have any concerns about our products,
you can contact us on
ProductSafety@springernature.com

In case Publisher is established outside the EU,
the EU authorized representative is:
Springer Nature Customer Service Center GmbH
Europaplatz 3, 69115 Heidelberg, Germany

Printed by Libri Plureos GmbH
in Hamburg, Germany